No. 1250
$16.95

DIGITAL
ELECTRONICS
TROUBLESHOOTING
BY JOSEPH J. CARR

TAB BOOKS Inc.
BLUE RIDGE SUMMIT, PA. 17214

FIRST EDITION

FIRST PRINTING—NOVEMBER 1980
SECOND PRINTING—MAY 1981

Library of Congress Cataloging in Publication Data

Carr, Joesph J
      Digital electronics troubleshooting.

      Includes index.
      1. Digital electronics. 2. Electronic apparatus
and appliances—Maintenance and repair. I. Title
TK7868.D5C36   621.3815   80-21138
ISBN 0-8306-9677-6
ISBN 0-8306-1250-5 (pbk.)

Contents

Introduction...8

1 Overview of Digital Electronics..9
What is Digital Electronics?—Digital Unmasked

2 Number Systems...15
Self-Evaluation Questions—Introduction to Number Systems—
Number System Conversions—Binary Arithmetic—
Recapitulation—Questions—Problems

3 Digital Codes...34
Self-Evaluation Questions—Hexadecimal Code—Split-Octal—
Binary-Coded Decimal—Excess-3 Code—Gray Code—
Alphanumeric Codes

4 Digital IC Logic Families.....................................46
What are Logic Families?—Resistor-Transistor Logic—Diode-
Transistor Logic—Transistor-Transistor Logic—Emitter-Coupled
Logic—High-Threshold Logic—Complementary Metal Oxide—
Semiconductor Logic—Schottky TTL ICs

5 Logic Gates...67
Self-Evaluation Questions—NOT Gates—Inverters—OR
Gates—AND Gates—NAND Gates—NOR Gates—Exclusive-OR
(XOR) Gates—Summary of Gate Actions—Experiments

6 Arithmetic Circuits..89
Self-Evaluation Questions—Adder Circuits—Subtractor Circuits

7 Flip-Flops: Clocked and Unclocked.........................96
Self-Evaluation Questions—R-S Flip-Flops—Clocked R-S Flip-
Flops—Master-Slave Flip-Flops—Type-D Flip-Flops—J-K Flip-
Flops

8 Some Advice 114
Poor Documentation—Something for TV Shops

9 Digital Counters: Devices and Circuits 117
Decimal Counters—Synchronous Counters—Preset Counters—Down and Up-Down Counters—Up/Down Counters—TTL/CMOS Examples

10 Display Devices and Decoders 134
Simply Binary Readout—N-Bit Binary Readout—Simple Decade Displays—Nixie Tubes—Seven-Segment Readouts—Decoders—Display Multiplexing

11 Registers 147
SISO and SIPO—Parallel—IC Examples

12 Timers and Multivibrators 152
Timers Versus Clocks—Long-Duration Timers—Programmable Timers

13 Data Multiplexers and Selectors 173
Multiplexing—IC Multiplexers/Demultiplexers

14 Data Transmission 188
Parallel Data Communications—Serial Data Circuits—Interface Standards—Teletypewriter Circuits—Universal Synchronous Receiver-Transmitter Chips—Tone Multiplexing—Telephone Lines

15 Computers 214
Microcomputers and Microprocessors—What is a Microprocessor?—What Can A Computer Do?—Types of Microcomputers—How Does a Computer Work?

16 The Z80—A Typical Microprocessor 229
Special-Purpose Registers—General-Purpose Registers

17 Memory-I/O Interfacing in Computers 234
Binary Word Decoders—Eight-Bit Word Decoders—I/O Select Circuits—Interfacing Memory—RAM R/W Memory Organization—Memory Bank Selection—Memory-Mapped I/O

18 Data Conversion 257
Data Converter Resolution—DAC Circuits—Analog-to-Digital Conversion

19 DC Power Supplies for Digital Equipment 287
Basic Principles of DC Power Supplies—Overvoltage Protection—Output Current Limiting—Typical Power-Supply Circuits—Typical Power-Supply Problems

20 Test Equipment for Digital Servicing 311
Logic Probes—Logic Analyzers—Tools

21 **Some Common Problems**.................................**320**
Power Line Problems—Other Glitches—Buses—RFI from Digital
Equipment—Uninterruptible Power Systems

22 **Other Equipment**.................................**335**
Printers—Paper Tape Readers—Magnetic Storage Devices—
Plotters/Recorders

Appendices

A— **Logic Level Detector Circuits**...............................**348**
B— **Four-Channel Oscilloscope Switch**.......................**349**
Index...**350**

Introduction

Digital electronics was once a very exclusive and limited field. Only a handful of technicians engaged in industrial, military, or computer electronics has to know anything at all about digital circuits. Even those few found that much of their daily work was, in reality, *analog* electronics rather than digital. Many industrial and military control systems, for example, were nothing more than dedicated, or preprogrammed, analog computers that performed a special function. But today, digital is everywhere. Computers, for example, are now cheaper, faster, and easier to operate than ever before. The early machines were dinosaurs compared with today's versions. My personal microcomputer, which easily fits on a desk top along with the typewriter used to type this book, has more computing power and is faster than the $156,000 computer of 10 years ago, yet is *150 times cheaper*.

Point of sale terminals are sprouting up all over the place. They will verify credit and checks and allow electronic transfer of funds from a customer's account to a merchant's. Bounced checks due to insufficient funds are becoming extinct.

And look at the consumer market. Microprocessor control of microwave ovens, home heating and air conditioning, TV and FM channel selectors, at least one car radio tuner, a large pile of electronic games and a host of other devices are digital.

In short, anyone in electronics at all is or soon will be servicing digital electronic circuits. Even technicians who have been in the trade for several decades must learn digital servicing. Younger technicians are finding more and more emphasis placed on digital electronics in their technical schools.

This is a self-study guide primarily aimed at technicians already in the field of electronics who want to learn somthing about digital circuits, possibly to save their job next year. It is also suitable for hobbyists, amateur radio operators, and technical school students who have already studied electronic basics.

Joseph J. Carr

Chapter 1
Overview of Digital Electronics

It is difficult to say where digital electronics began, but it is easy to say what it has become. Some engineers are fond of proclaiming that digital is the way of the future, and they may well be right. Although note well that it is a big mistake to count analog electronics as "down and out," because many electronics jobs are still best done by analog circuits. Many, perhaps most, transducers used in industrial applications, for example, are anlog in nature.

But despite the fact that analog is alive and well, digital has become a way of life in many areas of the electronics industry. People who once did not need to consider digital electronics at all, are now finding themselves immersed in it quite deeply. Many consumer electronic products, for example, now use digital circuits. Several television receivers use digitally-controlled phase-locked loop (PLL) tuners for VHF/UHF channel selection and other digital circuits to display data such as the channel number, date, or time of day on the cathode ray tube screen. One high-fidelity FM tuner on the market uses a microprocessor and a Delco digital car radio also uses PLL techniques. (see *Popular Electronics* for June, 1979). A stereo cassette deck on the market uses a microprocessor to control the tape drive system.

Tremendous growth has occurred in those areas of electronics that have traditionally seen a heavy digital involvement. Accelerated by the introduction of the microprocessor, this growth has caused some pundits and prognosticators to call it a new industrial revolution.

Look at the computer market as an example. From a handful of machines in the era immediately following World War II, the computer industry has grown to the point where every medium-sized business can afford a machine. Microcomputers can now do jobs for $5000 that only a decade ago required almost $100,000 worth of equipment! I find myself in awe, for example, of the fact that my own little desk-top microcomputer was put together for less than $1000 and is more powerful than the $50,000 machine used in engineering school during the late '60s.

The growth of microcomputers has helped to produce relatively sophisticated machines in a price range where doctors, dentists, lawyers, and other small business people can afford them. The system that now costs the same as a new car, once cost $150,000 and required an entire room to house it. Now, a simple desk top is sufficient. There has also grown up an industry of computer-related digital industries and industries that were not electronic at all until "digital" came along at cheap prices.

Products related to computers include cathode ray tube (CRT) video terminals, which are familiar to television repair technicians, and special point-of-sale computer terminals for retail establishments. The latter only superficially seem to appear like computer equipment and more nearly resemble calculators. The point-of-sale terminal may be little more than a special keyboard and a data communications device called a *modem* (*mo*dulator/*dem*odulator), or it may contain a real microprocessor device. Examples of these machines can be seen at almost any department store and supermarket cashier. Some grocery stores use special digital bar code readers and inventory control computers, and at least some fast food chains use special point-of-sale terminals in which the keys are labeled with the names of the food items offered. If an employee presses 3 and hamburger, the internal computer automatically totals up the price of three hamburgers, calculates and adds the tax, and then displays the price to the customer. These special terminals allow the merchant to use low-paid, unskilled labor in the retail operation, rather than more skilled people capable of operating the actual computer. What has worked for fast food chains also works in most other forms of mass merchandising. And it allows interrogation of credit account data and electronic transfer of funds, as well.

In the field of devices that were not previously digital—or even electronic—the electronic cash register has replaced the former mechanical monster. Also seen are produce/meat scales that are legal for trade and home bathroom scales that are not legal for trade but tell all too well if someone has been "illegal" on his diet. I recently bought gasoline in a self-service station that has a pump with an electronic digital readout—and an electronic ticket printer!

The upshot of all this activity is a terribly large increase in digital devices in the market place. Yet service capability has not kept pace with sales and introduction of new devices. The editor of a popular electronic servicing industry magazine (a technician's trade journal) once went to an electronics show and asked several displayers of electronic digital products who was going to service their wares. Their answers were all the same: television service technicians and others already in the business of electronic service. A number of these companies, especially the smaller ones, cannot afford to locate a service technician or field engineer within one hour's drive from every potential customer. Their only options are to require customers to send the product back to the factory or a regional service center, or appoint local servicers to do the job. Clearly, then, both market expansion and the inclusion of digital devices in previously nondigital or nonelectronic products creates a golden opportunity for people who want to get into digital servicing.

WHAT IS DIGITAL ELECTRONICS?

Digital electronic circuits differ from so-called analog circuits in that they are *binary* in nature; that is, they have only *two* possible states. These states are symbolically represented by the digits "0"

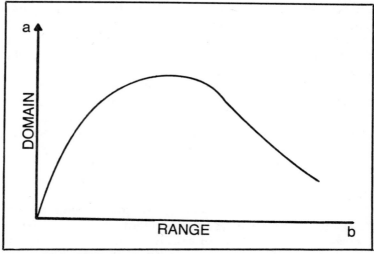

Fig. 1-1. Continuous analog signal.

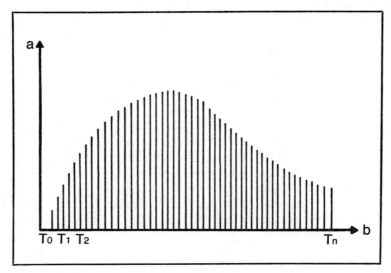

Fig. 1-2. Sample analog signal.

and "1" of the binary (base 2) number system, and they are actually two different voltage levels.

In most, but certainly not all, the 0 and 1 states represent voltage levels of 0 and +5 volts, respectively. These levels have become standard because they were used in the immensely popular transistor-transistor logic (TTL) integrated circuit devices. Modern complementary metal oxide semiconductor CMOS ICs, however, may use 0 and +5 volts, 0 and any positive potential between +4 and +18 volts, or a negative voltage for logical 0, and a positive voltage for logical 1 (−6 volts for logical 0 and +6 volts for logical 1 is popular).

Before proceeding further, however, let's first dispense with a little matter of notation. In order to avoid confusion, let us use the term LOW to denote logical 0, and HIGH to denote logical 1. This convention will be used every place in this book, unless we actually intend to denote the binary digits 1 and 0.

Let us further define digital signals by comparing three types of signals: *analog, sampled analog*, and *digital*. These are illustrated in Figs. 1-1 through 1-3.

The analog signal is shown in Fig. 1-1. Note that any signal can be defined as values within a *domain* (vertical axis) and a range (horizontal axis). In electrical circuits, these are *amplitude* and *time*, respectively. The analog signal is *continuous*, meaning that it can take on any amplitude from 0 to a, and it is defined at any instant in time over the interval 0 to b.

The analog signal, then, is defined for all values possible within the range and domain. This is not so with the sampled signal, however. It is defined for any amplitude from 0 to a but is only defined for certain *specific* points in time. The value of the amplitudes in Fig. 1-2 are known only at times T_0, T_1, T_2 ... T_n, but they are not known at instants between these times. This is not to say, however, that we cannot *infer* probable values at other instants from the overall shape of the curve. The method of inferring these unknown values is valid only if the sampling rate (number of samples per second) is greater that twice the highest frequency component present in the waveform being sampled.

A *digital* signal is defined for only specific values of amplitude and at specific instants in time. Furthermore, the amplitude values will be given in one of several binary codes (see Chapter 3) instead of actual voltage values. Although a graph is shown in Fig. 1-3, the binary values could just as easily be given in tabular form, as in Fig. 1-1D. Both forms convey the same information in binary systems.

Note that all three types of signal can denote the same information, but only the digital signal can be used in computers and other digital circuits. The other two forms must be first converted to digital form before being input to a digital instrument or computer.

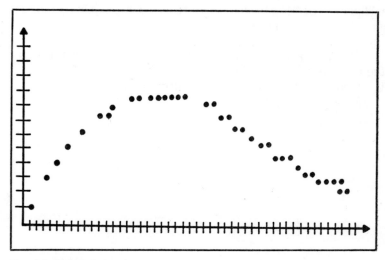

Fig. 1-3. Digitized signal.

DIGITAL UNMASKED

To many people whose sophistication is in nondigital areas of electronics, digital circuits often present something of a mystery. This was probably started, and is certainly sustained, by the computer industry. In reality, though, most digital electronic circuits are less difficult to understand than the workings of a color television receiver! Also, fewer technologies exist in digital equipment than in color TV. Digital equipment tends to use only digital electronics, while modern color TVs might use digital, VHF/UHF rf, pulse circuits, audio circuits, and a good mix of transistors and other discrete dvices, linear ICs, and digital ICs.

No one who is competent to service color TV receivers, two-way radios, or other electronic devices, need be afraid to service digital electronics. All of the concepts and devices are easy to understand and are imminently knowable.

Certain experiments in this book will aid you in learning the materials presented in the text. They are designed to give you a bit of "hands on" experience as you study the basics. Every effort has been made to select IC devices for these experiments that are widely available in hobby electronic stores, local electronic parts wholesalers, or mail-order electronic parts distributors. The latter may be found through their advertisements in publications such as *Popular Electronics, Radio-Electronics, QST, Byte, 73, Ham Radio*, etc. The experiments have also been designed around test equipment normally found in any properly equipped color television repair shop; no high-priced logic analyzers, or 200-MHz oscilloscopes, will be needed.

It is desirable that you buy a digital logic breadboard such as the Heath or the AP Products *Powerace 102*. The switch, pin, and LED indicator designations used in the experiments in this book are for the *Powerace 102*.

Chapter 2
Number Systems

You're going to learn a lot about various number systems in this chapter. Specifically, you'll learn about the binary number system, the notation of the octal and hexadecimal number systems, the principles of binary arithmetic, and the conversion of numbers in one system to equivalent numbers in another system.

SELF-EVALUATION QUESTIONS

These questions test your prior knowledge of the material in this chapter. Try answering them before you read the text. Then look for the answers as you read. When you are finished studying this chapter, try again to answer the questions. If you cannot answer a particular question, place a check mark beside it, and go back and reread appropriate portions of the text. When you are finished, try answering the questions at the end of the chapter using the same technique.

☐ Does $101_{10} = 101_2$?
☐ Convert $3F_{16}$ to octal notation.
☐ Binary, octal, decimal and hexadecimal are examples of number systems.
☐ What is the decimal equivalent of 362_8?
☐ Describe the procedure for binary subtraction using *twos complement* notation.

INTRODUCTION TO NUMBER SYSTEMS

One of the first complaints heard from new students of digital electronics is over the matter of learning the *binary* (base 2) number system. But this is highly desirable, and one step further, it is also essential that you become at least familiar with the *octal* (base 8) and *hexadecimal* (base 16) number systems. The two latter systems are frequently used in computer work as a shorthand for the binary numbers actually used in the computer. For example, binary 01001110_2 can also be equivalently represented as hexadecimal $4E_{16}$ or octal 116_8.

We all grew up using the *decimal*, or base 10, number system, so it is seemingly simpler to us. But it is, in reality, no easier than base 2, base 8, or base 16, or even base 99, for that matter. The same basic concepts used in decimal numbers are also found in binary, octal, and hexadecimal. Only the specific digits change between the number systems; the concepts and ideas remain the same. A comedian/mathematician from an early '60s television show used to claim that "base 8 is exactly like base 10 if you're missing two fingers!" While that statement was designed to raise laughs from parents of children struggling with "new math," it is none the less true. Similar statements could easily be made of the other bases. Of course, with hexadecimal, you would need *extra* fingers or use some of your toes.

Because several base number systems are used in this text, we will adopt a special notation. The base will be given as a subscript. Therefore, "101_2" means "101" in the binary (base 2) system, while "101_8" means "101" in the octal (base 8) number system, "101_{10}" is "101" in the decimal (base 10) system, and "101_{16}" is "101" in hexadecimal (base 16).

Decimal Numbers

Before tackling number systems in other bases, let us first review the basic concepts of all number systems, using the familiar base 10, or decimal, number system as an example. If you passed third grade, then you learned this system. And even if you did not get past the third grade, but are able to figure out if the clerk who sold you this book gave the correct change, then you still learned the decimal number system. After relearning the basics, you will tackle the binary, octal, and hexadecimal number systems.

All four number systems that we have mentioned (binary, octal, decimal, and hexadecimal) are examples of *weighted* number systems. The actual value of a number is dependent upon the *digits* used, such as 0, 1, 2 . . . , and their respective *position* with respect to each other.

The decimal, or base 10, number system has 10 digits: 0, 1, 2, 3, 4, 5, 6, 7, 8, and 9. These digits each represent different quantities, or values that are familiar to you already.

With only 10 digits, however, we can represent only 10 different quantities. But instead of creating a different digit—a digit is merely a *symbol*—to represent each possible quantity from zero to infinity, *positional notation* is used. This means that a digit's position relative to other digits confers added value. For example, the decimal number 682_{10} means:

$$682_{10} = (6 \times 10^2) + (8 \times 10^1) + (2 \times 10^0)$$

or

$$682_{10} = (600) + (80) + (2)$$

The base of the number system (10, in this case) is called the *radix*. In general, all weighted number systems use a *digit* multiplied by a *radix raised to some power* at each position. The general form for all weighted number systems is:

$$D_n R^n + \ldots + D_3 R^3 + D_2 R^2 = D_1 R^1 + D_0 R^0$$

where the D terms are the digits (0 through 9 in base 10), and the R terms are the radix (10 in base 10).
When we place our example, 682_{10}, in the general form, it is

$$(0 \times 10^n) + \ldots + (0 \times 10^3) + (6 \times 10^2) + (8 \times 10^1) + (2 \times 10^0)$$

Note: Any number raised to the *zero* power, such as 10^0, is equal to 1. So $10^0 = 2^0 = 8^0 = 16^0 = N^0 = 1$.

We can use the general form equation to represent either whole or partial quantities. A *partial quantity* is a fraction. A decimal point must be used to separate the two, such as:

$$+ \ldots + D_3 R^3 + D_2 R^2 + D_1 R^1 + D_0 R^0. \quad D_{-1} R^{-1} + D_{-2} R^{-2} +$$

$$D_{-3} R^{-3} + \ldots +$$

The digits to the left of the decimal point represent whole quantities, while those to the right represent fractional quantities.
For example, 18.23 means:
$$18.23_{10} = (1 \times 10^1) + (8 \times 10^0). + (2 \times 10^{-1}) + (3 \times 10^{-2}),$$

or

$$18.23_{10} = (10) + (8). (2 \times 0.1) + (3 \times 0.01)$$

$$18.23_{10} = 10 + 8 . + 0.2 + 0.03 = 18.23_{10}$$

Binary Numbers

The binary number system is just like the decimal number system if you're missing eight fingers. Such a situation would leave you two fingers, or two digits, which could be used for counting. These digits are 0 and 1. In the notation of the general form equation, then, a weighted binary number system would be

$$D_N 2^N + \ldots + D_3 2^3 + D_2 2^2 + D_1 2^1 + D_0 2^0$$

where the D terms are the digits (0 and 1 in binary).

For example, the binary number system number 01011_2 is the shorthand way of writing:

$$(0 \times 2^4) + (1 \times 2^3) + (0 \times 2^2) + (1 \times 2^1) + (1 \times 2^0) =$$

$$(0) + (8_{10}) + (0) + (2_{10}) + (1_{10}) = 10_{10}$$

At this point, we'll illustrate why the subscript notation must be used to let you know which system is being used. The digits "10" do not represent the same quantity in binary and decimal systems:

$$10_2 = (1 \times 2^1) + (0 \times 2^0) = 2_{10}$$
$$10_{10} = (1 \times 10^1) + (0 \times 10^0) = 10_{10}$$

Obviously, then, 10_2 does not equal 10_{10}.

The term *bit* used in digital electronics means *b*inary dig*it* and is the smallest unit of data possible. A bit can be either of the binary digits, 0 or 1. A *byte* is an array of *eight bits* arranged in a positional number representation from 00000000 to 11111111.

☐ Exercise 1-1:

Find the *decimal* values of the following binary numbers. Use the notation of $D_N 2^N + \ldots + D_3 2^3 + D_2 2^2 + D_1 2^1 + D_0 2^0$.
For instance,

Table 2-1. Digital Codes.

Decimal	Hexadecimal	Octal	Binary	Decimal	Hexadecimal	Octal	Binary
0	0	0	0	16	10	20	10000
1	1	1	1	17	11	21	10001
2	2	2	10	18	12	22	10010
3	3	3	11	19	13	23	10011
4	4	4	100	20	14	24	10100
5	5	5	101	21	15	25	10101
6	6	6	110	22	16	26	10110
7	7	7	111	23	17	27	10111
8	8	10	1000	24	18	30	11000
9	9	11	1001	25	19	31	11001
10	A	12	1010	26	1A	32	11010
11	B	13	1011	27	1B	33	11011
12	C	14	1100	28	1C	34	11100
13	D	15	1101	29	1D	35	11101
14	E	16	1110	30	1E	36	11110
15	F	17	1111	31	1F	37	11111
				32	20	40	100000
				33	21	41	100001

This procedure converts all numbers to decimal form:

1. Write down the number being converted.
2. Multiply the most significant digit (the left-most) by the radix of the number.
3. Add the result of step No. 2 to the next most significant digit (MSD).
4. Multiply the result obtained in step No. 3 by the radix.
5. Repeat steps 3 and 4 until all digits are exhausted. The final result is the answer.

□ **Example 2-1:**
Convert 1101101_2 to decimal form.

1101101	Multiply MSD by radix	$1 \times 2 = 2$
1101101	Add result to next digit	$2 + 1 = 3$
	Multiply result by radix	$3 \times 2 = 6$
1101101	Add result to next digit	$6 + 0 = 6$
	Multiply result by radix	$6 \times 2 = 12$
110111	Add result to next digit	$12 + 1 = 13$
	Multiply result by radix	$13 \times 2 = 26$
1101101	Add result to next digit	$26 + 1 = 27$
	Multiply result by radix	$27 \times 2 = 54$
1101101	Add result to next digit	$54 + 0 = 54$
	Multiply result by radix	$54 \times 2 = 108$
1101101	Add result to next digit	$108 + 1 = 109_{10}$

□ **Example 2-2:**
Find the decimal value of 234_8.

234_8	Multiply the MSD by radix	$2 \times 8 = 16$
234_8	Add result to next digit	$16 + 3 = 19$
	Multiply result by radix	$19 \times 8 = 152$
234_8	Add result to next digit	$152 + 4 = 156_{10}$

□ **Example 2-3:**
Find the decimal equivalent of $7F3A_{16}$.

$7F3A_{16}$	Multiply the MSD by the radix	$7 \times 16 = 112$
$7F3A_{16}$	Add result to next digit	$112 + 15 = 127$
	Multiply result by radix	$127 \times 16 = 2032$
$7F3A_{16}$	Add result to next digit	$2032 + 3 = 2035$
	Multiply result by radix	$2035 \times 16 = 32560$
$7F3A_{16}$	Add result to next digit	$32560 + 10 = 32570_{10}$

This procedure converts any number in any base, R, to the equivalent number in decimal notation. The procedure to follow does just the opposite and converts a decimal number to an equivalent in base R.

1. Divide the decimal number by the radix of the new number system.
2. The remainder from this step becomes the least significant digit (LSD) of the equivalent number in base R.
3. Divide the quotient from step No. 1 by the radix.
4. The remainder from this operation becomes the next digit of the new number system.
5. Continue this process until the digits are exhausted.

□Example 2-4:
Find the binary equivalent of 109_{10}.

Operation	Remainder
2 / 109	—
2 / 54	1 LSD
2 / 27	0
2 / 13	1
2 / 6	1
2 / 3	0
2 / 1	1
0	1 MSD

This result agrees with that of Example 2-1

$$109_{10} = 1101101_2$$

□ Example 2-5:
Convert 156_{10} to octal.

Operation	Remainder
8 / 156	4 LSD
8 / 19	3
8 / 2	2 MSD
0	

This result agrees with Example 2-2.

$$156_{10} = 234_8$$

□ Example 2-6:
Find the hexadecimal equivalent to $32,570_{10}$.

Operation	Remainder	
	Dec.	Hex.
16 / 32570	10	A LSD
16 / 2035	3	3
16 / 127	15	F
16 / 7	7	7 MSD
0		

This result agrees with Example 2-3.

$$32570_{10} = 7F3A_{16}$$

These procedures allow you to convert into decimal form from radix R, or from decimal into radix R. It is usually less trouble to use both procedures when converting from radix R1 to radix R2, when neither R1 nor R2 are decimal. For example, when converting from hex to octal, it may be better to convert from hex to decimal, and then decimal to octal.

□ **Example 2-7:**

Find the hex equivalent of 377_8.

1. *First* find the decimal equivalent of 377_8.

377_8	Multiply MSD by radix	$3 \times 8 = 24$
377_8	Add result to next digit	$24 + 7 = 31$
	Multiply result by radix	$31 \times 8 = 248$
377_8	Add next result to next digit	$248 + 7 = 255$

The decimal equivalent of octal 377_8 is 255_{10}.

2. Now, find the hexadecimal equivalent to 255_{10}.

16 / 255		
16 / 15	15	LSD
0	15	MSD

The hexadecimal notation for the digit "15_{10}" is F, so the answer is FF_{16}. We now know that the hexadecimal equivalent of 377_8 is FF_{16}:

$$377_8 = 255_{10} = FF_{16}$$

BINARY ARITHMETIC

Binary arithmetic is used in computers and is also very handy to use when trying to figure out almost any complex digital instrument. With more and more instruments on the market employing the microprocessor chip, the understanding of binary arithmetic techniques is even more critical.

Addition

The rules governing binary addition are simple; some people believe that they are simpler than decimal arithmetic! The rules are as follows:

$$0 + 0 = 0$$
$$1 + 0 = 1$$
$$0 + 1 = 1$$
$$1 + 1 = 0 \text{ plus carry } 1$$

☐ **Example 2-7:**
Add the binary numbers 01001_2 and 01110_2.

```
     1
   C\1 0 0 1
 + 0 |1 1 1 0
   ─────────
   1 0 1 1 0
```

☐ **Example 2-8:**
Add the binary numbers 00001_2 and 01001_2.

```
         1
   0 1 0 0\1
 + 0 0 0 0\1
   ─────────
   0 1 0 1 0
```

Subtraction

The rules for straight binary *subtraction* are as follows:

$$0 - 0 = 0$$
$$0 - 1 = 1 \text{ borrow } 1$$
$$1 - 0 = 0$$
$$1 - 1 = 0$$

Note, however, that it is difficult to build a digital electronic circuit that performs straight binary subtraction operations. It is, however, relatively trivial to build binary *adder* circuits. In most digital devices that are required to subtract, therefore, it is simpler to use *twos' complement* arithmetic to fool the circuit into thinking that it is adding!

Before discussing subtraction by twos' complement, let's find out what "twos" complement is. We begin by telling you about the ones' complement of a binary number. The complement of any binary number is its inverse; i.e., the complement of 1 is 0, and the complement of 0 is 1. To form the ones' complement of any binary number, therefore, we change all of the 1s to 0s and change all of the 0s to 1s.

Digit	Complement
0	1
1	0

For example, the complement of 1 0 1 1 0 0 1 1 is 0 1 0 0 1 1 0 0. Note that all ones became zeroes and all zeroes became ones.

□ Example 2-9:
Find the complement of 1011001_2.

number	1011001
complement	0100110

The twos' complement of any binary number is found by adding 1 to the ones' complement of that number.

□ Example 2-10:
Find the twos' complement of 1011001_2.

Number	1 0 1 1 0 0 1
Ones' complement	0 1 0 0 1 1 0
Add 1	+ 1
	0 1 0 0 1 1 1

0100111_2 is the twos' complement of 1011001_2.

Twos' Complement Subtraction

A binary subtrahend from a binary minuend by *adding* the twos' complement of the subtrahend to the minuend. Keep in mind that subtraction can be viewed as adding a negative number, and that twos' complement notation is a form of representation for negative binary numbers. The following are several examples of twos' complement arithmetic, given several different types of situations that could arise.

Before studying the examples, however, a standard convention concerning the use of "+" and "–" signs is important. An extra digit will be added, separated from the number by a comma (,), to denote the *sign* of the number. A 0 will represent positive (+) numbers, while a 1 will represent negative (–) numbers. See Fig. 2-1.

□ Example 2-11:
Perform subtraction when minuend and subtrahend are *both positive*, and the minuend is *larger* than the subtrahend.

Subtract $0,0100_2$ (4_{10}) from $0,0111_2$ (7_{10}).

Minuend	0,0111	+7
Subtrahend	− 0,0100	−(+4)
	?	?

1. Complement the	0,0100	
subtrahend	1,1011	+7
2. Add 1	+1	−4
	0,1100	?

3. Add twos' complement
   of subtrahend to the
   minuend

	0,0111	+7
	+1,1100	−4
1⟵	0,0011	3

(carry 1 discarded)

The problem is worked correctly because $0,0011_2 = +3_{10}$. Notice that the carry one is generated, but in this case it is discarded or ignored).

□ **Example 2-12:**

Perform subtraction of a *negative subtrahend* from a *positive minuend* when the minuend is *larger* than the subtrahend.

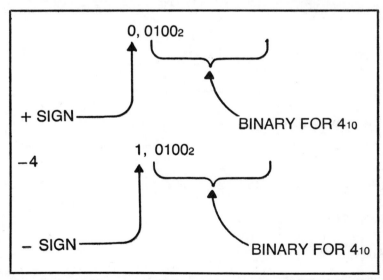

Fig. 2-1. The zero represents positive numbers, while the one represents negative numbers.

Subtract $1,0100_2$ (i.e. -4_{10}) from $0,0111_2$ (+7).

$$
\begin{array}{cc}
0,0111 & +7 \\
-1,0100 & -(-4) \\
\hline
? & ?
\end{array}
$$

Note in the decimal version that we are subtracting a minus quantity, so the problem is the same as adding their respective values:

$$
\begin{array}{cc}
0,0111 & +7 \\
+\,0,0100 & +4 \\
\hline
0,1011 & +11
\end{array}
$$

We know the problem is correctly worked because $0,1011_2 = +11_{10}$.

□ **Example 2-13:**

Perform subtraction of a positive subtrahend from a larger, negative minuend.

Subtract $0,0100_2$ $(+4_{10})$ from $1,1001_2$ (-9_{10}).

$$
\begin{array}{cc}
1,1001 & -9 \\
-0,0100 & -(+4) \\
\hline
? & ?
\end{array}
$$

Complement the *minuend* because it is a negative number.

$$
\begin{array}{l}
1,1001 \\
\text{becomes} \\
0,0110 \\
+1 \\
\hline
\end{array}
$$

Add 1 $0,0111$

This is the twos' complement of the minuend.

Add the twos' complement of the minuend to the subtrahend.

$$
\begin{array}{cc}
0,0111 & -9 \\
+0,0100 & +4 \\
\hline
0,1011 & -5
\end{array}
$$

Recomplement the answer

$$
\begin{array}{l}
1,0100 \\
+1 \\
\hline
1,0101
\end{array}
$$

Add 1

indicates negative

number ——————➤◀— binary for 5_{10}

The problem is correctly worked because $1,0101_2 = -5_{10}$.

Multiplication

Binary multiplication can be done using the same method as decimal multiplication, or by *repetitive addition*. The usual method by pencil and paper is shown in Example 2-14.

☐ **Example 2-14:**
Multiply 1101_2 (D_{16} or 13_{10}) by 0100_2 (4_{16} and 4_{10}).

$$
\begin{array}{r}
1101_2 \\
\times\ 0100_2 \\
\hline
0000 \\
0000 \\
1101 \\
0000 \\
\hline
0110100_2
\end{array}
\qquad
\begin{array}{r}
D_{16} \\
\times\ 4_{16} \\
\hline
34_{16}
\end{array}
\qquad
\begin{array}{r}
13_{10} \\
\times\ 4_{10} \\
\hline
52_{10}
\end{array}
$$

Multiplication is nothing more than addition performed over and over again. For example, 4×5 is

$$4_{10} \times 5_{10} = 4 + 4 + 4 + 4 + 4 = 20_{10}$$
$$\text{5 repetitions of } +4$$

The same thing can be done with binary numbers; the technique is called *repetitive addition* or the *shift left* method. The technique is:

1. Set the answer initially to zero; i.e., 00000000_2.
2. Write down the multiplication so that the least significant digit of the accumulated answer is aligned with the LSD of multiplicand.
3. Inspect, the multiplier beginning with the LSD. If it is a 1, then add the multiplicand to the answer. But if it is a 0, then add 0 to the answer.
4. Repeat steps 2 and 3 until finished.

Examine the process by performing the same problem as in Example 2-14 to see if we get the same answer.

☐ **Example 2-15:**
Use repetitive addition to solve Example 2-14.

$$
\begin{array}{ll}
1101_2 & \text{Multiplicand} \\
\times\ 0100_2 & \text{Multiplier} \\
\hline
? &
\end{array}
$$

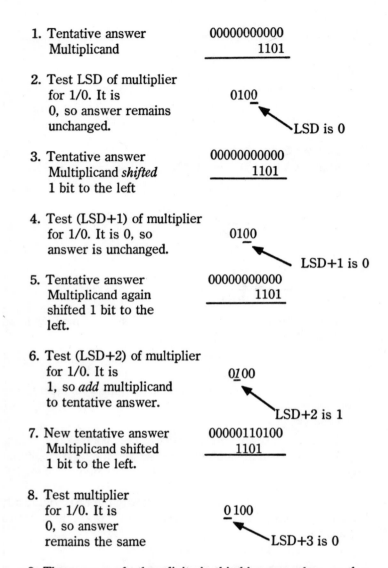

1. Tentative answer   00000000000
   Multiplicand   1101

2. Test LSD of multiplier
   for 1/0. It is   0100
   0, so answer remains
   unchanged.   LSD is 0

3. Tentative answer   00000000000
   Multiplicand *shifted*   1101
   1 bit to the left

4. Test (LSD+1) of multiplier
   for 1/0. It is 0, so   0100
   answer is unchanged.
                            LSD+1 is 0

5. Tentative answer   00000000000
   Multiplicand again   1101
   shifted 1 bit to the
   left.

6. Test (LSD+2) of multiplier
   for 1/0. It is   0100
   1, so *add* multiplicand
   to tentative answer.
                         LSD+2 is 1

7. New tentative answer   00000110100
   Multiplicand shifted   1101
   1 bit to the left.

8. Test multiplier
   for 1/0. It is   0 100
   0, so answer
   remains the same   LSD+3 is 0

9. There are no further digits in this binary number, so the answer is the same as in step 7: 00000110100.

Because the zeroes to the left of the most significant 1 are merely leading zeroes, drop them to shorten the answer to 110100.

The repetitive addition method seems a little cumbersome. On pencil and paper this is true, but in digital electronic circuits, adding and left-shifting are very easy to accomplish. In fact, it is

Table 2-2. Various Logical Operations.

NOT	NOT-A	$\overline{A}$	
AND	A AND B	$A \times B$*	
OR	A OR B	$A + B$	
XOR	A XOR B	$\underline{A + B}$	
NOR	A NOR B	$\overline{A + B}$	(A NOT-OR B)
NAND	A NAND B	$\overline{A \times B}$	(A NOT-AND B)

* Alternate form: A • B

NOT Function.   Def.: Output is the inverse of the input.

$$
\begin{array}{cc}
A & \overline{A} \\
0 & 1 \\
1 & 0
\end{array}
$$

NAND Function. $A \times B$
Def.: Output is 1 only if both inputs are 1.

$$
\begin{array}{rcl}
1 \times 1 & = & 1 \\
1 \times 0 & = & 0 \\
0 \times 1 & = & 0 \\
0 \times 0 & = & 0
\end{array}
$$

OR Function. $A + B$
Def.: Output is 1 if either input is 1

$$
\begin{array}{rcl}
1 \times 1 & = & 1 \\
1 \times 0 & = & 1 \\
0 \times 1 & = & 1 \\
0 \times 0 & = & 0
\end{array}
$$

Exclusive-OR (XOR) $A + B$

Def.: Output is 1 if either input is 1, but not if both
     inputs are 1.

$$
\begin{array}{rcl}
1 + 1 & = & 0 \\
0 + 1 & = & 1 \\
1 + 0 & = & 1 \\
0 + 0 & = & 0
\end{array}
$$

NOR Function. $\overline{A + B}$

Def.: Output is 0 if either input is 1.

$$
\begin{array}{rcl}
\overline{1 + 1} & = & 0 \\
\overline{0 + 1} & = & 0 \\
\overline{1 + 0} & = & 0 \\
\overline{0 + 0} & = & 1
\end{array}
$$

difficult to do other operations, where addition and shifting are duck soup.

Logical Operations

Digital circuits perform *logic operations* called NOT, OR, AND, and Exclusive-OR (XOR). The OR and AND functions also have inverted versions called NOR and NAND. In each case, two (or more) bits are compared, and the logical outcome is determined by which of these basic functions is being used.

Again consider some matters of notation. The NOT function is an inverting function. In the NOT function, a 1 will become 0, and a 0 will become 1. We represent the NOT function as a symbol with a bar overhead; i.e., $\overline{A}$ means the same as "NOT A." This means that if A = 1, then the $\overline{A}$ value is 0. If A = 0, then $\overline{A}$ = 1. The NOT symbol is also used in conjunction with other functions, as is the idea represented. A NOT gate connected with an OR gate produces a NOR gate. Similarly, a NOT gate connected with an AND gate produces a NAND gate.

Multiplication (×) and addition (+) symbols are used to represent the AND and OR functions, respectively. A circled addition symbol is used to represent the exclusive-OR (XOR) function. The NAND and NOR functions are represented by the equivalent AND and OR expressions with a NOT bar overhead. See Table 2-2.

RECAPITULATION

Now go back to the beginning of the chapter and try answering the self-evaluation questions. When you are finished, try answering the following questions and solving the following problems.

QUESTIONS

☐Define the term *bit*.
☐Define the term *byte*.
☐The octal number system is base—_____.
☐The decimal number system is base—_____.
☐The hexadecimal number system is base—____.
☐The binary number system is base—_____.
☐List all digits used in the binary number system.
☐List all digits used in the decimal number system.
☐List all the digits used in the octal number system.
☐List all digits used in the hexadecimal number system.
☐The *radix* of the binary number system is____.

☐List the rules for binary addition.
☐List the rules for binary subtraction.

PROBLEMS

☐Convert 101_2 to octal notation.
☐Convert $3F_{16}$ to octal notation.
☐Convert 110_2 to decimal.
☐Convert 029_{16} to decimal.
☐Convert 123_8 to decimal.
☐Perform the addition below

$$\begin{array}{r} 11011_2 \\ +101_2 \\ \hline \end{array}$$

☐Use twos' complement notation to perform the following binary subtraction.

$$\begin{array}{r} 1101_2 \\ -101_2 \\ \hline \end{array}$$

Chapter 3
Digital Codes

In this chapter, you'll learn the most common code schemes used in digital circuits to represent binary data. Also, you'll learn the types of transducers sometimes used to generate codes.

SELF-EVALUATION QUESTIONS

These questions test your prior knowledge of the material in this chapter. Try answering them before you read the text and then look for the answers as you read. When you have finished studying this chapter, try again to answer these questions. If you cannot answer a particular question, place a check mark beside it and go back and reread appropriate portions of the text. When you are finished, try answering the questions at the end of the chapter using the same technique.

☐ The decimal equivalent of BCD 0100 0110 1001 is _____.
☐ Represent hexadecimal number FF A4 in split-octal notation.
☐ The_____code changes only one bit for each increment.
☐ Three keyboard codes found in computer terminals are_____,_____, and_____.

A digital code is a binary representation of data. Nothing is mysterious about codes; they are merely methods for presenting data in formats that digital machines can easily use. A code may represent numerical, alphabetic, alphanumeric, control signals, etc. Most computer or teletypewriter keyboard codes, for example, contain numbers, alphabetic characters (upper and lower case), and control codes, such as carriage return.

One property of any useful machine code is that it is binary in nature, so that *on-off* switching circuits (discussed elsewhere in this book) can be used in the machine. It turns out that binary circuits are easiest to use in such cases.

Binary numbers are a form of "code" used to represent numerical quantities, but ordinarily, a binary code *does not* correspond to the binary number with the same bit structure. For example, in the ASCII code to be discussed shortly, the numerical character "7" is 00110111, while the binary number "7" (when expressed with the same number of bits) is 00000111. It is important to remember the difference between a *symbol* and a *number*. A computer may perform an arithmetic operation in which the result is the number seven (representing a quantity), but the output display device, which could be a printer or CRT video terminal, may require ASCII coding. If the programmer does not first convert the binary number 00000111 to 00110111, an incorrect result will be displayed to the outside world. In this particular example, it just happens that 00000111 in ASCII is the BEL symbol, so the signal bell of the printer would ring instead of character "7" being printed. It would be a rare situation, indeed, if the binary representation for a quantity was the same as the code for the character (s) that represents that quantity.

HEXADECIMAL CODE

A binary code of, say, four bits could be represented by a hexadecimal (base 16) code. This would greatly simplify entering data or instructions into a digital machine. Entering 7F, for example, is a lot easier than entering all eight binary digits, or 01111111, needed in the binary representation. Using hexadecimal notation permits four bits to be entered at a time (if eight bits is a byte, then are four bits a nybble?). Let us say that we have an eight-bit binary number:

$$1\ 0\ 1\ 1\ 1\ 1\ 0\ 1$$

To enter this number on a bit-by-tedious-bit basis would mean using eight separate keystrokes: 1-0-1-1-1-1-0-1. But we can break the number 101111101 into two four-bit groups:

$$1\ 0\ 1\ 1\qquad 1\ 1\ 0\ 1$$

Because 1011 is hexadecimal number B_{16} and 1101 is hex D_{16}, the hexadecimal number BD_{16} represents the binary pattern 10111101.

The hexadecimal system can also represent binary words that have bit lengths not divisible by four, provided that we assume the leading digits to be zeroes. For example, assume a ten-bit number:

$$1\ 0\ 0\ 0\ 1\ 0\ 1\ 0\ 1\ 1$$

Table 3-1. Binary Coded Decimal.

Decimal	BCD	Decimal	BCD
0	0000	5	0101
1	0001	6	0110
2	0010	7	0111
3	0011	8	1000
4	0100	9	1001

We can represent this with three hexadecimal digits:

$$0\ 0\ 1\ 0 \qquad 0\ 0\ 1\ 1 \qquad 1\ 0\ 1\ 1$$

Fill in the extra positions with zeroes. Because 0010 is hexadecimal 2_{16} and 1011 is hexadecimal B_{16}, hexadecimal $22B_{16}$ represents the same quantity as binary 1000101011 (no extra zeroes this time).

SPLIT-OCTAL

We can also represent binary data in the *split-octal* system by grouping the binary bits in groups of not more than three each. For example, the binary code 10111101_2 was represented by hexadecimal BD_{16} in the previous section. This same binary number can be divided up into octal groups:

$$010 \qquad 1\ 1\ 1 \qquad 1\ 0\ 1$$

Fill in any leading zeroes that are needed. Because binary 010 is octal 2_8, binary 111 is octal 7_8, and binary 101 is octal 5_8, octal 275 produces the same binary bit pattern as binary number 10111101. In this case, however, only three keystrokes (2-7-5) are needed instead of eight (1-0-1-1-1-1-0-1). Both octal and hexadecimal notation can represent the numerical, alphabetic, and control codes of code systems that are commonly used on keyboards, printers, teletypewriters, etc.

BINARY-CODED DECIMAL

The *binary-coded decimal* (BCD) is used to represent the 10 digits of the decimal number system in a four-bit format. See Table 3-1.

BCD is merely another form of decimal notation that is compatible with computers and other digital circuits. BCD words are therefore grouped exactly the same as regular decimal digits. The actual value of a BCD word is determined by its position relative to the other words, in a power of 10 system. For example, see Table 3-2.

Table 3-2. Decimal to Binary Coded Decimal Conversion.

Decimal	BCD
2	0010
23	00100011
175	000101110101

The BCD number system is difficult to use in making calculations, but it is not impossible. Its main use is in systems requiring a numerical output display, such as a frequency counter, digital panel meter, clock, scientific instrument, a toy, etc.

Several BCD decoder chips will convert a BCD input to an output capable of driving the principal display devices, which could be Nixie® tubes, seven-segment readouts, and so on. These are discussed in a later chapter.

EXCESS-3 CODE

The excess-3 code is formed by adding three to the BCD numbers. ($3_{10} = 011_2$). For example, see Table 3-3. Excess-3 code is used to make digital subtraction in BCD a little easier. Recall that complement arithmetic is often used with binary digits. This cannot be done in BCD, however, because it would occasionally result in a disallowed bit pattern that is not recognizable as any of the 10 BCD numbers.

GRAY CODE

Shaft encoders and certain applications work best if the binary code generated *changes only one bit at a time* for each change

Table 3-3. The Excess-3 Code.

	BCD	Excess-3
0	0000	0011
1	0001	0100
2	0010	0101
3	0011	0110
4	0100	0111
5	0101	1000
6	0110	1001
7	0111	1010
8	1000	1011
9	1001	1100

Table 3-4. The Gray Code.

Decimal	Binary	Gray Code
0	0000	0000
1	0001	0001
2	0010	0011
3	0011	0010
4	0100	0110
5	0101	0111
6	0110	0101
7	0111	0100
8	1000	1100
9	1001	1101
10	1010	1111

of state. An example of such a code is the Gray code, shown in Table 3-4. The Gray code is used most often where a transducer on a mechanical device is directly generating digital data, such as shaft position.

ALPHANUMERIC CODES

Actually, several different codes each represent alphabetic or numeric *characters*. Remember the distinction between characters and numbers—both represent quantities, but in different ways. We will discuss the obsolete but still found Baudot code, the popular ASCII code, and the IBM EBCDIC code.

Alphanumeric codes are used in cases where there is an output display device, such as a CRT video terminal, teletypewriter, or computer printer. Standardization, at least to a few different codes, allows connection of devices by different manufacturers.

Baudot Code

One of the earliest machine codes in general use was the Baudot code (see Table 3-5). Used in most of the earlier models of teletypewriter machines, Baudot has been largely replaced by another, more modern code (ASCII). Many Baudot machines are still in daily use; in fact, as this paragraph is being written, I can hear an old Model 27 Teletype® in the office across the hall! Amateur radio computer who favor radio teletype (RTTY) operation and amateur computer buffs on the lookout for a reliable, low-cost, hard-copy printer have kept the prices of these older machines high, even though it is a surplus market. Note that most teletypewriters will have a keyboard with a shift key just like a

Table 3-5. The Baudot Code.

B5	B4	B3	B2	B1	Regular	Shifted
0	0	0	0	0	Blank	Blank
0	0	0	0	1	E	3
0	0	0	1	0	linefeed	linefeed
0	0	0	1	1	A	-
0	0	1	0	0	space	space
0	0	1	0	1	S	Bell
0	0	1	1	0	I	8
0	0	1	1	1	U	7
0	1	0	0	0	Car. Ret.	Car. Ret.
0	1	0	0	1	D	$
0	1	0	1	0	R	4
0	1	0	1	1	J	'
0	1	1	0	0	N	,
0	1	1	0	1	F	!
0	1	1	1	0	C	:
0	1	1	1	1	K	(
1	0	0	0	0	T	5
1	0	0	0	1	Z	"
1	0	0	1	0	L	)
1	0	0	1	1	W	2
1	0	1	0	0	H	#
1	0	1	0	1	Y	6
1	0	1	1	0	P	0
1	0	1	1	1	Q	1
1	1	0	0	0	O	9
1	1	0	0	1	B	?
1	1	0	1	0	G	&
1	1	0	1	1	(figures)	(figures)
1	1	1	0	0	M	.
1	1	1	0	1	X	/
1	1	1	1	0	V	;
1	1	1	1	1	(letters)	(letters)

regular typewriter. In most cases, the capital letters are on the shifted keyboard and the lower case letters are on the unshifted keyboard. But some teletypewriters are just the opposite.

The Baudot code uses five bits, so its capable of representing only 2^5, or 32, different characters or control signals. A shift control, much like that found on a regular typewriter, allows another 32 characters, for a total of 64. Note that not all 64 characters are used on all Baudot-encoded teletypewriters.

ASCII Code

The American Standard Code for Information Interchange (ASCII) is the code most often used on computer and video terminal keyboards. It is especially common among hobby computers. The ASCII code is a seven-bit code and can represent up to 2^7, or 128, different characters or control signals. Recall that this is twice the capacity of the five-bit Baudot machines.

An eight-bit format is often used in ASCII machines, especially those used with eight-bit microcomputers. The eighth-bit is used as a parity indicator in some cases and as a strobe bit in others. A strobe bit, incidentally, tells the computer or other digital machine that the data on the seven data lines is valid and stable. Most machines are designed to ignore activity on the ASCII input line until a strobe bit is present.

In the ASCII listings shown in Table 3-6, the hexadecimal form of notation is used. The bit pattern for the specific character can be found by converting from hex to binary. An "&" sign is given in hexadecimal as 26_{16} and would be 0010 0110, or simply, 00100110_2.

EBCDIC Code

The Extended Binary Coded Decimal Interchange Code (EBCDIC) is most often associated with IBM Corporation equipment. This code can be most easily explained by referring to the Hollerith card shown in Fig. 3-1. This card was originally used to automate the tabulation of the 1890 United States census, yet remains intensely popular today, It is, in fact, one of the mainstays of the data processing industry.

The Hollerith card is divided into rows and columns. There are 12 horizontal rows and 80 vertical columns. There are also two areas on this card. The rows numbered 0 through 9 form one area, while rows 11 and 12 form the other. Punches made in rows 11 and 12 are called *zone punches*.

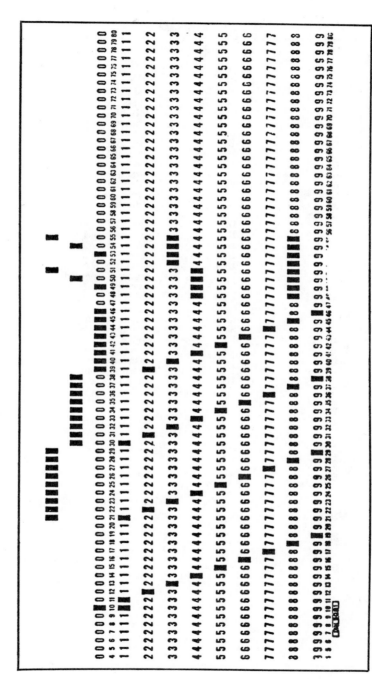

Fig. 3-1. Hollerith card.

41

Table 3-6. ASCII Code.

Hex Code	Meaning	Comments
00	NUL	null
01	SOH	start of heading
02	STX	start text
03	ETX	end text
04	EOT	end of transmission
05	ENQ	enquiry
06	ACK	acknowledgement
07	BEL	bell
08	BS	back space
09	HT	horizontal tab
0A	LF	line feed
0B	VT	vertical tab
0C	FF	form feed
0D	CR	carriage return
0E	SO	shift out
0F	SI	shift in
10	DLE	data link escape
11	DC1	direct control 1
12	DC2	direct control 2
13	DC3	direct control 3
14	DC4	direct control 4
15	NAK	negative acknowledgement
16	SYN	synchronous idle
17	ETB	end of transmission block
18	CAN	cancel
19	EM	end of medium
1A	SUB	substitute
1B	ESC	escape
1C	FS	form separator
1D	GS	group separator
1E	RS	record separator
1F	US	unit separator
20	(special)	—
21	!	—
22	"	—
23	#	—
24	$	—
25	%	—
26	&	—
27	'	—
28	(	—
29	)	—
2A	*	—
2B	+	—
2C	,	—
2D	-	—
2E	.	—
2F	/	—
30	0	—
31	1	—
32	2	—
33	3	—
34	4	—
35	5	—
36	6	—
37	7	—
38	8	—
39	9	—

Hex Code	Meaning	Comments
3A	:	—
3B	;	—
3C	>	—
3D	=	—
3E	<	—
3F	?	—
40	@	—
41	A	—
42	B	—
43	C	—
44	D	—
45	E	—
46	F	—
47	G	—
48	H	—
49	I	—
4A	J	—
4B	K	—
4C	L	—
4D	M	—
4E	N	—
4F	O	—
50	P	—
51	Q	—
52	R	—
53	S	—
54	T	—
55	U	—
56	V	—
57	W	—
58	X	—
59	Y	—
5A	Z	—
5B	[	—
5C	\	—
5D	]	—
5E		—
5F	i	—
60	∧	—
61	a	—
62	b	—
63	c	—
64	d	—
65	e	—
66	f	—
67	g	—
68	h	—
69	i	—
6A	j	—
6B	k	—
6C	l	—
6D	m	—
6E	n	—
6F	o	—
70	p	—
71	q	—
72	r	—

Table 3-6. ASCII Code.

Hex Code	Meaning	Comments
73	s	–
74	t	–
75	u	–
76	v	–
77	w	–
78	x	–
79	y	–
7A	z	–
7B	{	–
7C			–
7D	}	–
7E	~	–
7F	DEL	–

Characters are represented by one-byte binary words. For this discussion, we can divide the byte into two *nybbles* of four bits each. This convention allows us to use hexadecimal notation. The code for digits is shown in Table 3-7.

Conversion to binary merely requires the substitution of the binary equivalents of the hexadecimal numbers. For example:

Decimal	EBCDIC	Binary
6	F6	11110110

EBCDIC representation of numbers places a hexadecimal F_{16} in the most significant position, so the four most significant binary digits will be 1111_2. Those positions are used when representing alphabetic characters and indicate which third of the alphabet the letter falls in. There are 26 letters in our alphabet, so the alphabet can be divided roughly into thirds. See Table 3-8.

Two punches are required to represent a letter on a Hollerith card. A zone punch tells us which third of the alphabet in which the letter appears, and the numeric punch tells us the letter number. Letters in the first third of the alphabet have a row-12 zone punch, those in the second third of the alphabet have a row-11 zone punch, while those in the third third of the alphabet have a row-∅ zone punch.

Table 3-7. Hexadecimal Notation.

Digit	Hex Code	Digit	Hex Code
0	F∅	5	F5
1	F1	6	F6
2	F2	7	F7
3	F3	8	F8
4	F4	9	F9

Table 3-8. EBCDIC Representation of the Alphabet.

Letter No.	1st Third	2nd Third	3rd Third
1	A	J	-
2	B	K	S
3	C	L	T
4	D	M	U
5	E	N	V
6	F	O	W
7	G	P	X
8	H	Q	Y
9	I	R	Z

Consider, for example, the letter D. It is the fourth letter in the first third of the alphabet. It will be represented by a row-12 zone punch and a row-4 numeric punch, both in the same column. The letter N, on the other hand, is the fifth letter in the second third of the alphabet. It is, therefore, represented by a row-11 zone punch and a row-5 numeric punch.

Table 3-9. Alphanumeric Portion of the EBCDIC Code.

Character	Hex Code	Character	Hex Code
A	C1	O	D6
B	C2	P	D7
C	C3	Q	D8
D	C4	R	D9
E	C5	S	E2
F	C6	T	E3
G	C7	U	E4
H	C8	V	E5
I	C9	W	E6
J	D1	X	E7
L	D2	Y	E8
M	D3	Z	E9
N	D4		

A slight departure from the system is noted in the last third of the alphabet. There is no row-1 punch. The first letter in this group is S and it is represented by a row-0 zone punch and a row-2 numeric punch. The alphanumeric portion of the EBCDIC code is shown in Table 3-9.

Chapter 4
Digital IC Logic Families

Integrated circuits are not all that difficult. In fact, after this chapter you'll have learned the different types of digital integrated circuits and how these ICs are used. Also, you'll learn the problems associated with each IC family and how they are avoided and the special handling procedures for certain ICs that are sensitive to static electricity damage.

WHAT ARE LOGIC FAMILIES?

A logic family is a group of IC devices manufactured using similar technology to possess similar attributes and work together using direct interconnection rather than elaborate interfacing circuitry.

Several different families and subfamilies have been developed over the past several decades: resistor-transistor logic (RTL), diode-transistor logic (DTL), transistor-transistor logic (TTL), emitter-coupled (AC) logic (ECL), high-threshold logic (HTL), complementary metal oxide semiconductor (CMOS) logic, and Schottky TTL.

Some of these types of device are considered obsolete, while others are still in current production. We will cover the obsolete devices briefly, however, because they may still be found in some older equipment, even though no longer incorporated into new designs.

One of the aspects of digital electronics that is amusing to many observers is that each of the presently used logic families has its own staunch adherents who vigorously defend their favorites. Those who like CMOS above all others, for example, are forever pointing out the virtues of CMOS and the failures of TTL. They usually succeed in failing to mention the shortcomings of CMOS. The "TTL-niks," on the other hand, exalt the positive properties of TTL and completely fail to mention TTL problems. The truth is that each of the currently popular families has its own place, and electronic circuit designers must make decisions based on what they preceive as the reasonable trade-offs between the vices and virtues of the respective families. One young fellow I know boasted to his very experienced electronics instructor that he could "do anything in CMOS that can be done in TTL, without all of that high current drain!" The instructor coolly asked, "Can you build an 80-MHz counter or shift register?" The answer is a resounding NO, because CMOS is a slow-speed family.

Speed and *power consumption* tend to be directly proportional to each other in digital devices for good reason. High speed, then, usually means increased power consumption. It seems that low power operation, with its reduced current drain, means higher internal resistances and usually higher internal capacitances. This makes the RC time constants of the circuit longer, thereby accounting for the slower speeds. When you reduce the impedances inside the device and the interelectrode capacitances in order to make the device faster, such as for operation at a higher frequency, then you also increase the current drain; hence, you increase the power consumption.

Before proceeding to a discussion of the various families of IC device, let's first dispense with a little housekeeping. What is meant exactly by the terms *current source, current sink,* and *fan-out/fan-in*?

A current source is a circuit that will supply current to the load. If you connect a resistor to one side of a power supply, then it becomes a current source to circuits connected between the free end of the resistor and the other side of the power supply.

A current sink, on the other hand, is a circuit that accepts current from another circuit. In the simplified example above, the load connected to the current source will act as a current sink.

Most IC logic families have inputs and outptus that are matched as to their current source or sink roles. A TTL input, for example, acts as a current *source*. The TTL output, on the other hand, acts as a current *sink*. When an output is used to drive an input, therefore, the current produced at the input will find a path to ground through the output of the driving device.

No matter how else you view them, IC digital devices are still little more than collections of transistors, diodes and, resistors.

Just as in any direct-coupled transistor circuit, one must pay attention to the drive levels, impedance matching, and voltage/current levels at the interface between two stages. In most analog electronic circuits, the designer must do this each and every time a new circuit is designed, but the digital electronics industry has devised a method for specifying input drive requirements and output drive capabilities. The popular TTL family, for example, has standardized on a +5 VDC power supply for all devices. An output might produce as much as +5 VDC, and all input terminals must be capable of accepting that level. Similarly, the properties of a standard TTL input are specified as to current source level in milliamperes. The basic unit used in this spec, then, is the TTL input. With this in mind, the terms *fan-out* and *fan-in* are understandable. The fan-out is an integer number, applied to the output of a device, that tells the designer how many TTL inputs it can drive. If the rating is a "fan-out of 10," the designer knows that the device will drive up to 10 standard TTL input loads. The fan-in is the same type of number, but specifies how many TTL input loads this particular input requires. Most devices have a fan-in of 1 on each input. The designer merely adds up the total fan-in at a point and makes sure that the driving device has a fan-out equal to or greater than this number.

RESISTOR-TRANSISTOR LOGIC

One of the earliest logic families used in IC form was *resistor-transistor logic* (RTL). This family uses type numbers in the μL900 range and 700 range. Two examples are the μL900, which is a popular NAND gate, and the 789, which is a J-K flip-flop.

In most logic families, you get a reasonable picture of typical internal circuitry by examining an inverter, or simple gate, circuit. Figure 4-1 shows an RTL inverter stage. The RTL inverter IC would contain all three components: R1, R2, and Q1.

This circuit should be familiar to all readers, as it is merely a version of the simple common-emitter amplifier circuit using an NPN transistor. Of course, in digital circuits, transistor Q1 is operated in the saturated mode. When a +3.6V potential is applied to the collector, and the ground terminal is connected to the negative side of the power supply, the stage will operate as an inverter. When there is no potential applied to the input, or when the input is grounded, transistor Q1 is cut off. No collector current will flow in Q1, so the output terminal will be at the supply potential (+3.6VDC). Apply a HIGH (+3.6V) potential to the

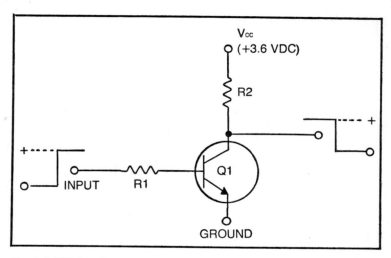

Fig. 4-1. RTL inverter.

input, however, and Q1 will saturate. This causes the collector to drop to within a few millivolts of ground potential. The output, then, is effectively grounded (a LOW condition). The requirement for inversion is met: A HIGH on the input creates a LOW on the output, and a LOW on the input creates a HIGH on the output.

RTL devices typically operate from a +3.6V power supply, and it was always considered imprudent to operate at higher potentials. A maximum of about +4V is usually considered the rule of thumb. On the positive side of the ledger, however, as long as voltages on a printed circuit board were all less than +4VDC, no combination of shorts or opens could destroy the IC. This is something that cannot be said of the other, more modern, logic families. TTL, for example, is easily destroyed by a slipped screwdriver or careless soldering iron.

RTL is considered to be a current-source family. Note that the output is connected to the +3.6VDC power supply through a resistor, which is similar to the situation in the example of the last section. When the output is HIGH, it will source current to the output through the resistor to the +3.6VDC supply.

DIODE-TRANSISTOR LOGIC

Diode-transistor-logic (DTL) devices are also obsolete, but are still found in equipment now in service. Most DTL devices bear 900-series type numbers.

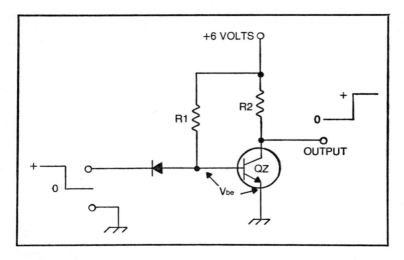

Fig. 4-2. DTL inverter.

Figure 4-2 shows a DTL inverter circuit. The integrated circuit device would contain all components shown, such as R1, R2, Q1, and D1. Most of these devices operated at DC potentials of +5 to +6 volts. The operation is as follows:

☐ When the input is HIGH (+6V), diode D1 is reverse biased. The base of Q1, then, is forward biased, so Q1 will conduct collector current. The output voltage at this time is zero.

☐ When the input is LOW (grounded or at 0V), D1 is forward biased, dropping V_{be} enough to reverse bias Q1. This condition causes the output voltage to go HIGH.

In some DTL circuits, HIGH is +5 to +6 volts, and LOW was 0-volts. In other cases, HIGH and LOW are ±6 volts. DTL operating speeds are approximately the same as RTL.

TRANSISTOR-TRANSISTOR LOGIC

Transistor-transistor logic (TTL) has been the most popular IC logic family for many years—so much so, in fact, as to be called ubiquitous. These devices are given type numbers in the 5400/54000 and 7400/74000 series. The 5400/54000 series devices are able to meet military specifications, while the 7400/74000 devices are the so-called "commercial" grade ICs. Devices with similar type numbers in these series are identical in function, if not specifications. A 5490, for example, is the same decade counter as a 7490.

TTL devices operate from a +5 VDC power supply. In many TTL devices, the power supply voltage must be maintained within a small range around +5V. The generally given range for all TTL devices is +4.75 to +5.2 VDC. Performance is unspecified at below the lower limit, and damage to the device can be expected at above the upper limit.

The increased operating speed (15 to 20 MHz is typical) is achieved at the expense of increased current consumption. The typical TTL device requires 15 to 30 milliamperes. If 22 mA is taken as "average," the current requirements rapidly add up. Fewer than 50 integrated circuits will draw 1A of current from the +5 VDC power supply. Many digital circuits, incidentally, have many more than 50 IC devices! Add to that the fact that optimum circuit performance often requires a power supply capable of delivering a lot more than "just enough" current, and you will see why high-current, 5V power supplies are frequently identified with TTL instruments.

Typical TTL inverter circuits are similar to Figs. 4-3 and 4-4. The device shown in Fig. 4-3 has a typical TTL output, while the circuit of in Fig. 4-4 has an open-collector output stage. The open-collector circuit requires an external pull-up resistor, but this allows the device to interface directly with lamps, light-emitting diodes, relay coils, and so forth.

The regular TTL output circuit shown in Fig. 4-3 uses a totem pole stage; transistors Q3 and Q4 form a series pair. Transistor Q2

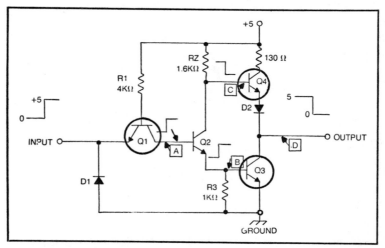

Fig. 4-3. Normal TTL inverter.

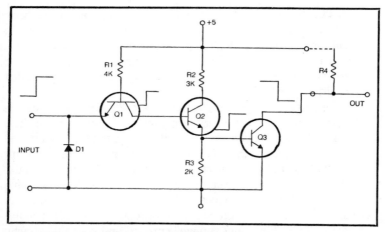

Fig. 4-4. TTL inverter with open-collector output.

forms a driver for Q3/Q4, while Q1 forms the input stage. Notice that the input of a TTL device is a transistor *emitter* terminal. The circuit or device, which could be another TTL IC device, that is connected to the input must be able to *sink* the emitter current of Q1. It can be said, then, that TTL inputs are current sources, while TTL outputs are current sinks. The operation of Fig. 4-3 follows.

When the Input is LOW:

☐ The emitter of Q1 is grounded, so Q1 is forward-biased.

☐ Resistor R1 and the *c-b* junction of transistor Q2 form the collector load of transistor Q1. Point A, which is where the collector of Q1 and the base of Q2 are connected together, is also LOW.

☐ When point A is LOW, Q2 is cut off, making point B LOW and point C HIGH.

☐ Under this condition, Q3 is turned off and Q4 is turned on, connecting the +5V supply to the output terminal.

When the Input is HIGH:

☐ Transistor Q1 is turned off, so point A is HIGH.

☐ When point A is HIGH, transistor Q2 is turned on, making point B HIGH and point C LOW.

☐ When point C is LOW, Q4 is turned off, disconnecting the +5 VDC supply from the output. Point B is HIGH, so Q3 is turned on. This causes the output terminal to be at ground potential.

An inverter circuit such as Fig. 4-3 needs no external components to operate. But that feature also makes it more difficult to interface the device with anything other than transistors or other TTL devices. In many cases, therefore, an *open-collector* IC is needed. This type of TTL inverter is shown in Fig. 4-4. Notice that the input and driver stages (Q1/Q2) are essentially the same as in the previous case. But only Q3 is used in the output stage, and it has no collector circuit. The collector of Q3 is brought to the outside world through a terminal and must be returned to a positive voltage through external resistor R4. The supply end of R4 must go to +5 VDC in some cases, while it may go to any voltage between +5V and either +15V or +30V in other cases, depending upon type number. An open-collector TTL IC would be used to turn on a lamp, LED, relay, and so forth.

The multiple input gates of the TTL series are similar to the circuits shown previously but will use a multiple emitter input stage (Fig. 4-5). Each TTL input must be a transistor emitter terminal.

EMITTER-COUPLED LOGIC

Emitter-coupled logic (ECL) is a high-speed IC logic family of nonsaturated devices. TTL, on the other hand, operates with the transistors in saturation or cutoff all of the time. Operating frequencies to 120 MHz are common. Some special and very costly devices are able to operate to frequencies in the *gigahertz*, or 1000 MHz, range. Many TTL frequency counters use ECL input stages to divide the input frequency down to TTL ranges, such as 50 to 80 MHz maximum.

Figure 4-6 shows a TTL gate circuit. This particular device is a four-input OR/NOR gate. Note that both inverted (NOR) and noninverted (OR) output terminals are available.

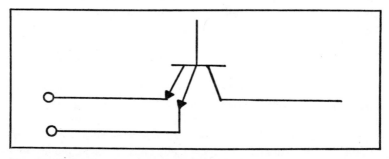

Fig. 4-5. Multiple inputs are created with extra emitters.

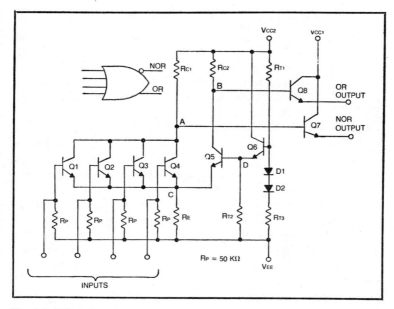

Fig. 4-6. ECL gate.

These devices offer optimum AC performance when V_{cc} and V_{ee} are set to ±5 VDC, respectively. The ECL family combines high speed with moderate power consumption, something usually deemed impossible in the saturated logic families (RTL, DTL, TTL, CMOS, etc.).

HIGH-THRESHOLD LOGIC

The TTL logic devices that have become almost the standards in the electronics industry have only 400 to 500 millivolts of *noise immunity*—the ability to not respond to noise potentials. In many applications, this is not sufficient to eliminate false operation. An industrial site that uses lots of electrically operated machinery will be considered electrically noisy. Pulses picked up from the electrical activity will be treated as a valid pulse by TTL and DTL logic elements.

The *high-threshold logic* (HTL), also called *high-noise-immunity logic* (HNIL), offers superior immunity to spurious noise pulses. The first HTL devices were designed to operate from 12V power supplies, while later devices operate from 15V power supplies. CMOS devices are technically high threshold but are not included in this discussion because they are sufficiently different to

54

warrant a separate section. Bipolar HTL devices have approximately 10 times the noise immunity of TTL. The improved noise immunity is derived from three factors: lower circuit impedances, slower operating speeds, and higher threshold potentials.

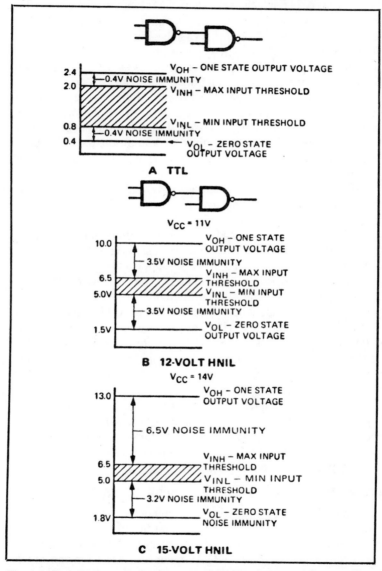

Fig. 4-7. Comparisons of noise immunities of HTL and TTL devices.

Additionally, HTL devices usually offer a ±1V power supply immunity, as opposed to the ±250mV tolerance of TTL. As a result, some circuits using HTL devices do not require regulation of the DC power supply.

Figure 4-7 shows the relative noise immunity of the two HTL families—12V and 15V—with the popular TTL family. Notice that TTL devices have a 400-mV (0.4V) noise immunity band. The 12V HTL/HNIL family has a 3.5V noise immunity band, almost 10 times better. The 15V family, however, does even better, having a 3.2V band at the lower end and 6.5V at the upper end.

Figure 4-8 shows a typical HTL gate. Notice that it is in some respects similar to the DTL circuit design, except for 5.8V Zener diode in series with the input diodes. The Zener diode operates to raise the voltage threshold required to operate the gate. An input signal must be high enough in level to exceed the Zener breakdown potential, or the gate will not respond.

HTL inputs are current sources, so any device driving an HTL input must be able to sink a current equal to approximately 2.1 mA in 12V devices and 2.6 mA in 15V devices.

HTL/HNIL devices can be interfaced with elements of other digital logic families. They can be interfaced directly with RTL/DTL/TTL inputs, *provided* that the V_{cc} supply of the HTL device is the same potential as the RTL/DTL/TTL device. Interfacing with HTL devices operated at other potentials, however, may require some external components. To interface an HTL/HNIL output with a TTL input, an HTL/HNIL device that has an open collector output is usually used. A 3K to 4K pull-up resistor can then be connected between the open-collector output terminal on the HTL/HNIL device to the +5 VDC power supply of the TTL device.

Similarly, TTL open-collector outputs can be interfaced with HTL/HNIL inputs if a pull-up resistor (10K ohm) to the +12V or +15V power supply is provided. A restriction on this latter application, however, is that the TTL open-collector must be a +15V or +30V type, such as a 7406 inverter.

COMPLEMENTARY METAL OXIDE SEMICONDUCTOR LOGIC

Complementary metal oxide semiconductor (CMOS) logic elements are among the most recent on the market. It is a little premature, however, to write the obituary of the earlier families, especially TTL and ECL. CMOS devices differ from the other families in that they use MOS field-effect transistors instead of

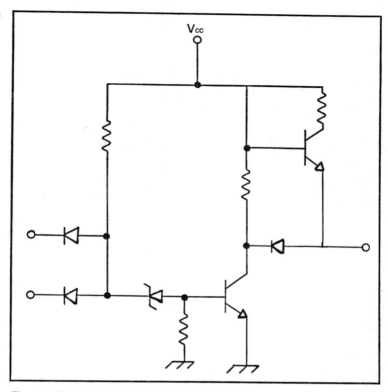

Fig. 4-8. Typical HTL gate.

bipolar (NPN and PNP) transistors. It is a saturated logic family that is well regarded for its low-current consumption.

It takes two types of MOSFETs to make a CMOS inverter stage: *p-channel* and *n-channel*. The p-channel device will turn on when its gate potential is zero with respect to the source, while the n-channel turns on when the gate is positive with respect to the source.

Figure 4-9 shows a basic CMOS inverter circuit that is designed with an n-channel and a p-channel MOSFET connected in series. The output is taken from their mutual junction—the source end of one and the drain end of the other. The gates are connected in parallel, so the same potential is applied to both.

When the input signal is LOW (0 volts or some minus potential—both can be used in any given CMOS device), transistor Q1 is turned off and Q2 is turned on. This allows the V_{cc} potential to appear at the output terminal. Only a very small resistor, such as

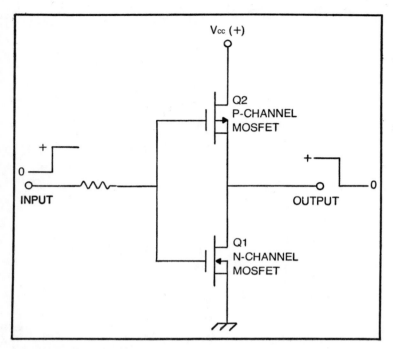

Fig. 4-9. CMOS inverter.

the channel resistance of Q2, is in series with the V_{cc} terminal and the output terminal.

When the input signal goes HIGH, on the other hand, exactly the opposite situation occurs; Q1 is turned on and Q2 is off. In this condition, there is a high resistance between the output terminal and V_{cc} and a low resistance between the output terminal and ground or the minus power supply.

A model of this action is shown in Fig. 4-10, using resistors and the DC supply. The resistances represent the channel resistances of Q1 and Q2 under the two different conditions.

In Fig. 4-10A the input is LOW, in which case Q1 is off and Q2 is on. This makes R_{Q1}, the channel resistance of transistor Q1, HIGH and R_{Q2} LOW. Because of voltage divider action between R_{Q1} and R_{Q2}, the output will be HIGH.

The reverse situation occurs, however, when the input is HIGH (see Fig. 4-10B). In this case Q1 is on and Q2 is off. This will reverse the relationship between the respective channel resistances of the two transistors.

The models shown in Fig. 4-10 demonstrate the really unique nature of CMOS devices. These devices will have a low resistance path to V_{cc} in the HIGH input state and a similar path to ground in the LOW input state. The off transistor in any given state will have a high-channel resistance, while the on transistor has a low channel resistance. This situation—a high and a low resistance in series—causes the power supply to always see a high impedance to ground. The only appreciable current drain is when the output stage is in *transition* from one state to another. At those times, the respective on-off roles of the transistors are changing to the opposite state. At

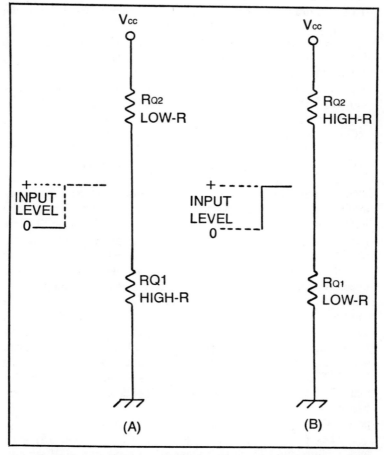

Fig. 4-10. Relative resistances of Q1/Q1 when input is LOW at A, and relative resistances of Q1/Q2 when input is HIGH at B.

sometime during this transition, the channel resistances will be lower than maximum and will be equal. The current at this instant is maximum. During the stable periods the current is in the microampere range.

How do the two most popular IC logic families compare with each other? Both CMOS and TTL devices have their own adherents who will swear *by* them—and *at* the other family. In truth, though, it must be said that both types of devices have their own virtues and in some areas the proper design selection is made by a coin toss; little or no difference would be noted.

When using the TTL device, be ready to accept the fact that TTL ICs draw a considerable amount of current from the power supply. Additionally, there is the fact that the TTL device must operate from a regulated +5V supply that is negative-grounded. No other supply will do. CMOS, on the other hand, will draw very small amounts of current (microamperes versus milliamperes), and is able to operate with a wide variety of power supply voltages. In many cases, the power supply for CMOS devices need not be regulated. Some designers take advantage that the CMOS device will operate from positive and negative supplies.

The difference in current requirements between CMOS and TTL is, perhaps, best illustrated by two similar gates. A certain TTL NAND gate requires 8 mA when the gate output is HIGH and 22 mA when the output is LOW. A CMOS package that is functionally identical, such as a NAND gate, requires only 15 mA in the HIGH condition and 160 mA in the LOW condition. It is this low power drain feature that makes CMOS very popular, especially in portable equipment. "Credit card" sized pocket calculators and digital wrist watches would not be possible if TTL devices were used! A digital watch uses only a few dozen microamperes in CMOS technology, but you would need to pull a wagon behind you to carry the batteries if both portability and long operation were required of the same design in TTL. In large-scale digital equipment, very heavy power supplies are needed when TTL devices are used. My own homebuilt microcomputer, for example, uses a lot of TTL and bipolar memory devices (both high current demand), so a 20A power supply at 5V is needed. Some large mainframe computers, the so-called "dinosaurs" (note the microcomputerist's bias), may draw up to 150A to 200A at +5V.

The operating voltage range of the CMOS devices is very wide, while the TTL device must see a regulated +5V supply. Most CMOS devices will operate properly over the range +4.5V to

+18V. In many cases, ±15 volts can be used. The minus power supply is used instead of the needed ground connection when monopolar power supplies are used; however, most CMOS devices are designed for optimum performance in the 7V to 12V range. Some complex, special function CMOS ICs will not operate properly at less than +7 VDC.

In contrast to the wide operating range of the CMOS device, we find that the TTL device will not operate well if the power supply voltage deviates very far from the nominal +5V value specified. In fact, many devices are listed as requiring a voltage tightly regulated between +4.7V and +5.2V. It is the experience of many, however, that many circuits will not operate at less than about 4.9V. This is especially true of the more complex special function devices, or where there are a lot of ICs in the circuits. Monostable multivibrators and clock circuits especially tend to become troublesome outside of the +4.9V to +5.2V range.

The high current drain requirements of TTL devices often cause problems in power-supply distribution on printed circuit (PC) boards. The tracks used to carry the +5V and ground return currents will sometimes cause *IR* drops sufficient to reduce the voltage at a chip below the operating range. In addition to this current starvation, there is also the problem of *glitches*, or spurious pulses in the circuit. These raise the dickens with gates, counters flip-flops, and monostables. A glitch usually consists of a sharp, negative-going pulse on the +5 VDC power supply line. It is caused when a sudden current demand occasioned by turning on a chip reduces the terminal voltage at that chip. Any PC board that contains even a modest number of TTL devices requires bypass capacitors (0.001 μF to 1 μF) generously sprinkled all over the board. There should be at least one 0.001-μF capacitor for each TTL package, and it should be located close to the package +5V line. Disc ceramic and tantalum bypass capacitors are usually specified for this application.

The glitch problem also forces us to use PC boards wherever possible. Regular wiring, increasingly popular on circuit boards due to the availability of low-cost wire-wrapping apparatus, often creates problems due to its inherent inductance. The inductance of a short piece of wire may not seem like much, but it becomes significant when the rise time is very fast, as it often is in TTL circuits, and the current at peak is high. Wire connections should be used only with caution in TTL circuits.

We also find that, in many cases, the noise immunity of the

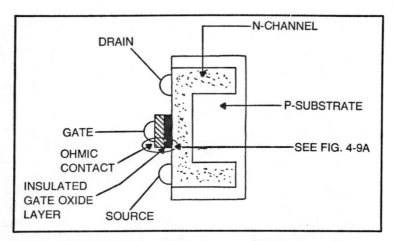

Fig. 4-11. MOSFET device showing breach in insulation from static.

CMOS device is superior to taht of the TTL. The TTL device will have but a few hundred millivolts of noise immunity, while the CMOS device has several volts. The reason for this difference is that the CMOS device will not change state unless the input signal passes the $\frac{1}{2}V_{cc}$ point. If a +12V supply is used, for example, the trip point differentiating between HIGH and LOW conditions is 12/2, or 6 volts. The TTL device, on the other hand, is not firmly in the LOW state unless the signal level is less than 800 mV (0.8 volts), and is not in the HIGH state unless the signal has a level over 2.4 volts.

CMOS is a slow logic family while TTL is fast. Almost any garden-variety TTL device will operate to 18 MHz, and most will operate to 20 MHz. Some selected devices or special type numbers operate to 80 MHz. On the other hand, CMOS operates only to 10 MHz at best, and most devices are limited to less than 5 MHz.

Another problem with the CMOS device is that it will be damaged from static electricity potentials that inevitably build up on your body, clothing, tools, instruements and the work bench itself! While most CMOS devices are safe to handle when connected into their circuits, they become exceptionally vulnerable when handled out-of-circuit. This does not, incidentally, give you license to nanдie the CMOS circuit card carelessly. Damage can and does occur in-circuit; it is merely less probable, not impossible. You will hear some technicians and engineers say that the problem is overstated by both manufacturers and technical writers. After having blown a few terribly expensive devices and caused damage to low-cost devices that were difficult to locate, I can only say, "More power to them." If they want to be the ones who blow it, then let them pay the price. As for me, I will try to use the handling procedures that follow.

The CMOS device is sensitive for the same reason MOSFET transistors are sensitive: low gate-channel breakdown voltage, which is typically 80 volts. The gate element of a MOSFET device is insulated from the channel by a layer of oxide insulating material that may be as thin as 0.0001 inch. That's 1/10,000! The voltage level of the static electricity on your body is often several hundred volts and may well exceed a kilovolt. This accounts for the spark and slight shock received when you discharge the static into a grounded object.

Figures 4-11 and 4-12 show the actual mechanism of CMOS/MOSFET failure. When the integrity of the insulating layer is breached, the metal ions of the gate element are drawn into the hole, creating an ohmic path between the channel and the gate. Unfortunately, this problem may not manifest itself immediately. The breach may not be large enough at first to cause an immediate short circuit, but after several weeks of operation there are enough metallic ions in the breach to cause a short. This may account for the disgusting habit some devices have of failing after a few weeks of operation—of course, well within the repair warrantee period you offered. The cure for this problem is to use special handling procedures and a special working environment in which everything is at the same potential—usually ground.

Some CMOS devices, such as the inverter circuit of Fig. 4-13, have internal Zener diodes that prevent the breakdown. These diodes will clamp the interelement voltages to some value less

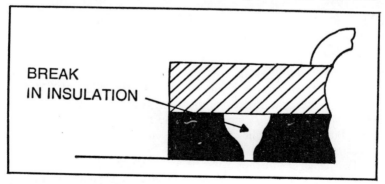

Fig. 4-12. Enlarged view of breached area.

than breakdown. These are not perfect, however, and cannot be trusted totally. The B-series CMOS devices, such as 4011B, do have this internal protection.

CMOS devices are shipped from the manufacturer in antistatic containers. They will be in black, plastic bags or on black foam. Some are stored in special containers that are not opaque, but are still antistatic; an ohmmeter reveals that the plastic is conductive. Do not remove CMOS devices from their shipping container or the foam backing until you are ready to use them.

The working environment that you should establish when working on CMOS circuits is a grounded environment. This will allow the static electricity to drain to ground before it can do any damage. One tactic is to use a large layer of aluminum sheet, metallic foil, or even an old cookie sheet as a work surface. This surface must be at ground potential. Similarly, your body and all of your tools and instruments must be grounded. The ground to your body can be an alligator clip attached to your metallic watchband, or it could be a more formal device made from an ID bracelet with a wire soldered to the flat surface.

Caution: It can be *deadly* and *dangerous* to keep your body at ground potential if there is any possibility of a high current flowing through it. Two different problems exist. Low-voltage, high-current power supplies will cause a third degree burn that in all probability will require medical treatment if that bracelet contacts the source. Secondly, if that death bracelet touches a 110 VAC source, a lethal AC current will probably flow. Contacting any point of your body with AC will probably be lethal if the environment is directly grounded. Do not ground the worksheet directly. Instead, connect a 1 megohm, or higher, resistor between your work surface and actual ground. A grounded work surface could make you a 5000W fuse and is a real widow(er) maker! Using a 1-megohm resistor still allows the surface to be maintained at ground potential but places a limit on the current that can flow. *Please stay alive.*

CMOS Handling Rules

Only a few rules are needed for safe handling of CMOS devices. Note that these same rules generally apply to the Schottky TTL devices discussed in the next section. Failure to follow these rules might cause you to build a "silicon to carbon converter."

☐ Always handle the CMOS devices, or printed circuit boards containing CMOS devices, in an environment grounded to earth through a 1-megohm, or greater, resistor.

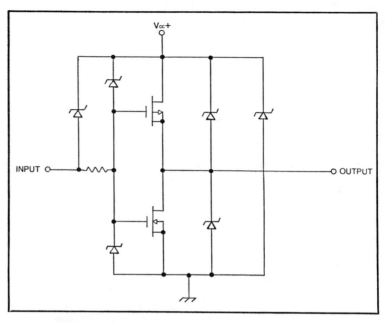

Fig. 4-13. Zener diode gate protection in CMOS gate.

☐ Keep CMOS devices or PC boards containing CMOS devices in their antistatic shipping containers until ready for use.

☐ When removing the CMOS device or PC board containing CMOS devices from its shipping container, ground yourself first through a 1-megohm resistor. Before actually touching the device or the PC board, touch the container or other conductive shipping surface. This equalizes the pin voltages.

☐ Always avoid touching the IC pins.

☐ Use only soldering devices that have grounded tips—those that have a three-wire AC power cord.

☐Turn off the power before inserting CMOS devices or PC boards containing CMOS devices into their sockets.

☐Avoid wearing synthetic fabrics.

☐Use test equipment that has a three-wire power cord; this indicates that the equipment cabinet is grounded.

☐ Handle PC boards containing CMOS devices only by their edges and avoid touching the exposed card-edge connector.

SCHOTTKY TTL ICS

Schottky TTL devices are a modified TTL family that uses a Schottky barrier diode (SBD) connected between the collector and

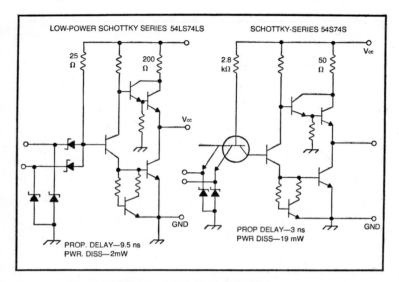

Fig. 4-14. Schottky and low-power Schottky inverters.

base of the logic transistors. This technique serves to reduce the base saturation, thereby reducing the charge stored in the transistor. A smaller stored charge means faster switching times.

Figure 4-14 shows an example of two common types of Schottky TTL digital IC logic elements. The low-power Schottky devices are designated by 54LS and 74LS type numbers; i.e., a 74LS121 is a low-power Schottky version of the 74121 TTL device. An "LS" gate will dissipate only 2 mW, with less than 10 nS of propagation delay time.

The regular Schottky devices are designated by 54S and 74S type numbers. They have only 3 nS of propagation delay time, but they require almost 20 mW of power from the DC supply. The "S" and "LS" devices may often be found mixed with regular TTL and are used to reduce system power requirements or enhance speed.

Chapter 5
Logic Gates

After reading this chapter, you'll know the different types of digital logic gates and the different gate ICs available in CMOS and TTL. Additionally, the operation of principal types of gates is covered.

SELF-EVALUATION QUESTIONS

☐ What happens to the output if one input of a 7400 gate is grounded?

☐ In a NAND gate, the output is____when both inputs are HIGH.

☐ Both inputs of a two-input XOR gate are HIGH. What is the condition of the output terminal?

☐ One input of a two-input NOR gate is HIGH. What is the condition of the output terminal?

Several elementary types of *logic gates* form the basic building blocks out of which even the most complex digital circuits are formed. Even the most complicated computers are merely large scale combinations of a few basic types of logic gate. Some of the digital ICs that will become very familiar to you in the near future as you study this text are merely combinations of different gate circuits in a seemingly monolithic IC form. All complex digital logic functions are made from very elementary gate types.

The types of gates which we will discuss are NOT, OR, AND, NAND, NOR, and Exclusive-OR (XOR). Various examples of these gates are found in each of the major IC logic families.

NOT GATES—INVERTERS

Figure 5-1 shows the symbol and truth table for the most elementary type of gate. This gate is called an inverter, com-

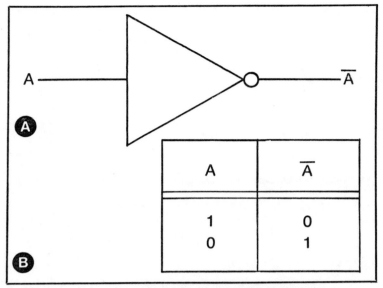

Fig. 5-1. Symbol for inverter at A, and truth table for inverter at B.

plementor, or NOT-gate. This type of stage will invert the input signal; that is to say the output of the inverter will be the opposite of the input. If the input is HIGH, then the output will be LOW. Similarly, if the input is LOW, then the output will be HIGH. Such gates are called *complementors* because, in digital circuits, HIGH and LOW levels are considered to be complements of each other.

We use the NOT terminology because of the symbols used denote input and output. If the input is called *A*, then the output will be called *not-A* because it will always be the logic state that *A* is not. The symbol used for not-A is "$\overline{A}$", sometimes called *bar-A*. NOT-A, however, is the preferred expression.

The truth table for a NOT gate is as follows:

A	A
1	Ø
Ø	1

Whenever A is at logic level 1 (HIGH), then the output $\overline{A}$ will be at logic level Ø (LOW). When A goes to level Ø, then $\overline{A}$ goes to 1.

The symbol for an inverter, or NOT gate, as shown in Fig. 5-1, is a triangle with a circle at the output. Anytime you see a digital logic element in a schematic that has a circle at the output, the output is inverted. Similarly, some multiple-input devices have a circle on one or more inputs. This symbol means that the logic function *for that input only* is inverted.

You will often see a triangle without a circle on the output. This is the schematic representation for *noninverting buffers*. Such devices will produce an output that is identical to the input but is usually capable of delivering more current than ordinary TTL outputs. Buffers are used to isolate one circuit from another, or to increase the drive capability (fan-out) of a circuit. Whenever a logic element must drive a long transmission line or a circuit that is heavily loaded with many inputs, a buffer might be required. Figures 5-2 and 5-3 show popular examples of digital inverters from the TTL and CMOS logic families.

The TTL 7404 device is shown in Fig. 5-2. This chip is called a *hex inverter* because it contains six independent TTL inverters. The positive side of the +5 VDC power supply is applied to pin No. 14, while the ground side is applied to pin No. 7. Although by no means universal, this pinout configuration for the power terminals is very common in digital IC devices. Note that there is no interaction between the inverter sections, so all six may be used independently. The average current drain is 12 to 15 mA.

Three related devices, using the same pinouts, are the 7405, 7406, and 7416. All three of these are *open-collector* chips, meaning that the output transistor requires an external pull-up resistor to the V_{cc} + power supply. The resistor connects the collector of the transistor, connected internally to the IC output terminal, to its power source. The 7405 is very similar to the 7404, except for the required 2.2K ohm pull-up resistor to the +5V supply. The 7405 draws 12 to 15 mA.

The 7406 and 7416 are open-collector devices, like the 7405, but are able to withstand 30V (7406) and 15V (7416), respectively. Note that the package voltage applied to pin No. 14 *must* remain at

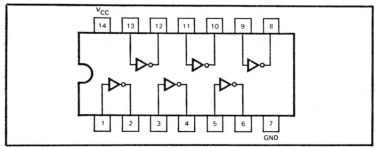

Fig. 5-2. TTL Hex inverter.

+5V; only the output transistor collector can operate at the higher potential. The higher collector voltage is intended to make it possible for the IC to drive an external load.

The 7406 device can sink up to 30 mA, while the 7416 can sink up to 40 mA, through the output transistor collector. Ths specification means that the 7406 load must have a DC resistance not less than R = (30V)/(0.03A), or 1000 ohms. The 7416 load must be not less than R = (15V)/(0.04A), or 375 ohms. These values refer to the dc resistance between the open-collector output and the V_{cc}+ power supply. Both 7406 and 7416 devices are frequently used as drivers for relay coils, LEDs, and incandescent lamps.

Figure 5-3 shows the pinouts for the 4009, 4049, and 4069 CMOS hex inverter chips. The 4009 is now considered to be obsolete for new designs and has been replaced by the 4049 device. The 4009 device uses two V_{cc}+ terminals and will be destroyed if these voltages are applied in the wrong sequence. The voltage applied to pin No. 1 must always be equal to or greater than the voltage applied to pin No. 16. Again, all six inverters may be used independently.

The 4049 device may be used as a level translator between CMOS devices and TTL devices. The voltage applied to pin No. 1 defines the output voltage swing, so if +5 VDC is used, the device becomes TTL-compatible. The 4049 can handle 3.2 mA, so it will drive two standard TTL loads. The input voltages applied to the 4049 inverter sections can swing to +15 VDC, regardless of the V_{cc}+ voltage applied to pin No. 1.

Although some people call the 4069 device a "low-power" version of the 4049, it is not. The 4069 has different pinouts and is not suitable for either direct interfacing with TTL logic or level translation of any other type. The 4069 will only interface directly with other CMOS devices.

The difference between the current drains of TTL and CMOS devices can be seen by comparing the 7404 and 4069 devices. At +5 VDC (used by all TTL devices), the 4069 draws 0.5 mA, as opposed to 15 mA for the 7404; this is a difference of 3000 percent! It is little wonder that designers of battey-powered devices prefer CMOS over TTL.

OR GATES

An OR gate will produce a logical-1 or HIGH output if *any* input is HIGH. Figure 5-4A shows the symbol for an OR gate, Fig. 5-4B shows an equivalent circuit that produces the same action, and Fig. 5-4C shows the truth table for the OR gate.

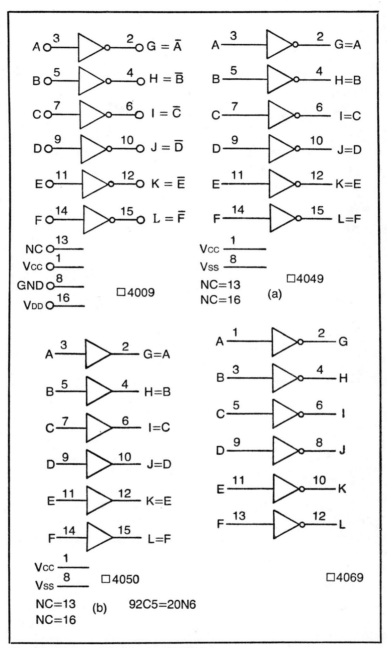

Fig. 5-3. Hex inverters from the CMOS line.

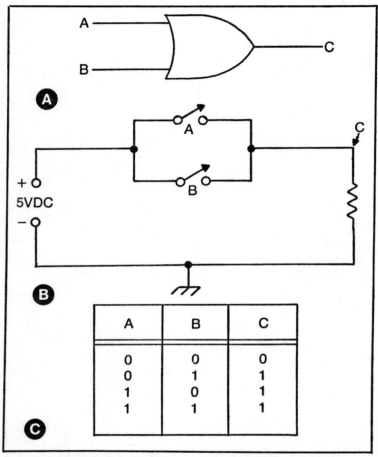

Fig. 5-4. OR gate symbol at A, electrical circuit model for OR gate at B, and truth table for OR gates at C.

Let us consider Fig. 5-4B in the light of the definition of an OR gate. Let the voltage at point "C" represent the output of the OR gate, and swtiches A and B the inputs of the OR gate. The logic is as follows:

Switch Condition	Input logic level	Output Condition
open	Ø	LOW
closed	1	HIGH

By the definition of an OR gate, we would expect C to be HIGH when *either* input is HIGH also (switch A OR switch B is closed), and LOW only when *both* inputs are also LOW (switch A and switch B are open). This is exactly the action found in the circuit of Fig. 5-4B. Voltage C goes to +5 VDC if either switch is closed and will be zero only when both switches are open. Ths condition is reflected in the truth table shown in Fig. 5-4C: C is HIGH if A OR B is also HIGH.

Examples of TTL and CMOS OR gates are shown in Figs. 5-5 and 5-6, respectively. The type 7432 TTL device is a quad, two-input OR gate. Note that the V_{cc}+ and ground terminals are pins 14 and 7, respectively. This is the same power supply configuration as found on the 7404 (discussed in the previous section). The 7432 requires approximately 20 mA of current per package.

A CMOS example, the 4071 device, is shown in Fig. 5-6. This IC is also a quad, two-input, OR gate. The current requirement is also 0.5 mA, and the difference between TTL and CMOS is seen even more clearly. But power consumption also brings speed. The propagation time—the time between input and the output response that it causes—is 12 ns for TTL, and 80 ns for CMOS.

AND GATES

The AND gate produces a HIGH output if *both* inputs are HIGH and a LOW input if *either* input is LOW. Figure 5-7A shows the circuit symbol for an AND gate, while Fig. 5-7B shows an equivalent circuit that does the same job, and Fig. 5-7C shows the AND gate truth table. In this type of circuit we obtain a HIGH output only if A AND B are also HIGH. There are few examples of TTL AND gates, but in the CMOS line we find the 4073, 4081, and 4082 devices. The 4073 is a triple, three-input AND gate. The 4081 is a quad, two-input AND gate, and the 4082 a dual, four-input device.

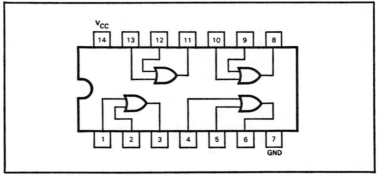

Fig. 5-5. TTL OR gate.

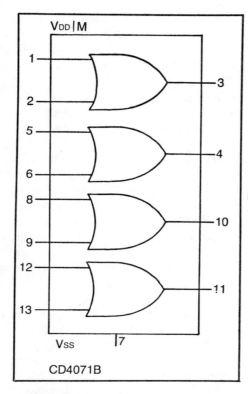

Fig. 5-6. CMOS OR gate type 4071.

CD4071B

NAND GATES

A NAND gate can be viewed as an AND gate with an inverted output; in fact, you can simulate the NAND gate by connecting the input of an inverter (NOT gate) to the output of an AND gate. The designation "NAND" is derived from *NOT-AND*. We sometimes see the NAND expression written in the form $\overline{AND}$, $\overline{AXB}$, or $\overline{A.B}$.

Figure 5-8 shows the NAND gate. Figure 5-8A shows the circuit symbol used in schematic diagrams. Figure 5-8B is an equivalent circuit that performs the same job, and Fig. 5-8C is the NAND gate truth table. The rules of operation for the NAND gate are as follows:

☐ A LOW on *either* input creates a HIGH output.
☐ A HIGH on *both* inputs is required for a LOW output.

We can see this same action in the equivalent circuit shown in Fig. 5-8B. Recall our switch logic protocol: Switch open is a LOW on the input, and a switch closed in a HIGH on the input.

74

The voltage at point "C" will be HIGH if either switch A or switch B is open (either input is LOW). But if both switches are closed (both inputs are HIGH), the load is shorted out and point "C" is LOW.

The NAND gate is one of the most popular gates used in digital equipment. An example of a quad, two-input, TTL NAND

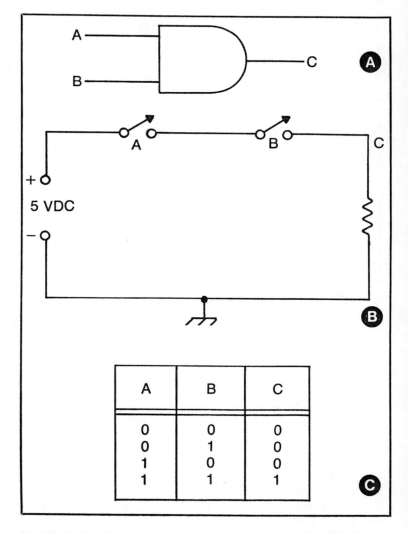

Fig. 5-7. AND gate symbol at A, electrical circuit model for an AND gate at B, and truth table for an AND gate at C.

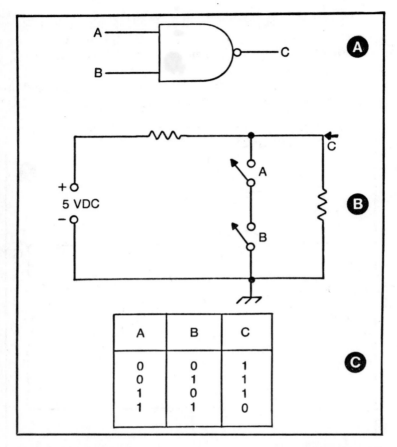

Fig. 5-8. NAND gate symbol at A, electrical circuit model for an NAND gate at B, and truth table for a NAND gate.

gate is shown in Fig. 5-9, the popular 7400 device. It is literally possible to construct circuits to replace any of the other gates, using only 7400 devices, although such would be an unnecessary and uneconomic course of action.

Recall that TTL is a current-sinking logic family, meaning that the output will sink, or pass to ground, current. The inputs of a TTL device form a 1.6-mA current source. Any TTL input is considered HIGH when greater than +2.4V is applied, and must be held to less than 0.8V when LOW. This requirement means that any ground connection must have a resistance less than R = (0.8 V)/(0.0018 A), or 440 ohms. If a higher ground resistance is encountered, or if a voltage between 0.8V and 2.4V is applied, then

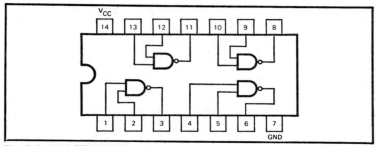

Fig. 5-9. 7400 TTL NAND gate.

operation is undefined—and all bets as to operation are off! The 7400 devices requires approximately 12 mA of current per package.

A CMOS NAND gate IC, the 4011, is shown in Fig. 5-10. This device is also a quad, two-input, NAND gate. THe 4011 requires

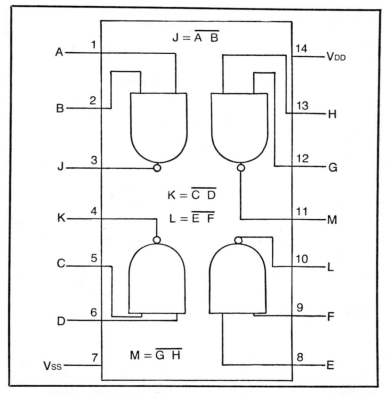

Fig. 5-10. CMOS NAND gate.

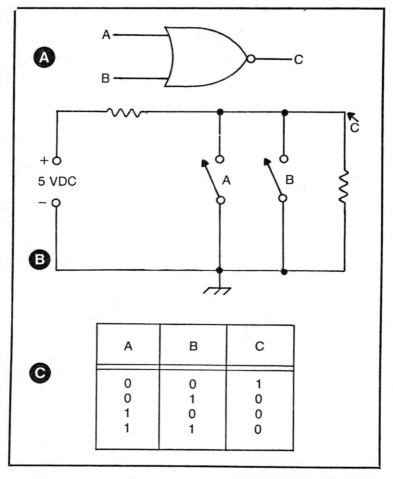

Fig. 5-11. Symbol for a NOR gate at A, electrical circuit model for a NOR gate at B, and truth table for a NOR gate at C.

only 0.4 mA per package, again a substantial savings over TTL. Various combination circuits using NAND gates will be discussed in subsequent chapters.

NOR GATES

A NOR gate is the same as an OR gate with an inverted output. The NOR designation is derived from "NOT-OR," and the NOR functions can be symbolized by $\overline{A + B} = C$. The circuit diagram symbol for the NOR gate is shown in Fig. 5-11A, while the equivalent circuit is shown in Fig. 5-11B. A truth table for the NOR gate function is shown in Fig. 5-11C. The rules governing the operation of the NOR gate are:

☐ A HIGH on *either* input produces a LOW output.
☐ If *both* inputs are LOW, then the output is HIGH.

We can see this action in the circuit of Fig. 5-11B. If either switch A or switch B is closed, then output C will be LOW. But when both switches are open, then output C is HIGH.

Figure 5-12 shows an example of a popular TTL 7402 NOR gate. This device contains four two-input gates that may be used independently of each other. The package draws about 12 mA.

EXCLUSIVE-OR (XOR) GATES

The exclusive-OR, or XOR as it is designated, is shown in Fig. 5-13. Examine the truth table for the XOR gate below:

A	B	C
Ø	Ø	Ø
1	Ø	1
Ø	1	1
1	1	Ø

From this truth table we may infer the rules for operation governing the exclusive-OR gate:

☐ The input is LOW if *both* inputs are LOW or if *both* inputs are HIGH.
☐ The output is HIGH if either input is HIGH.

In other words, a HIGH on either input *alone* produces a HIGH output, but a HIGH on both inputs simultaneously produces a LOW

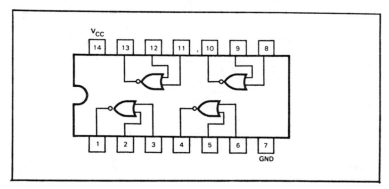

Fig. 5-12. 7402 TTL NOR gate.

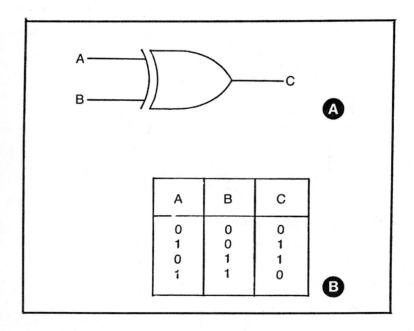

A	B	C
0	0	0
1	0	1
0	1	1
1	1	0

Fig. 5-13. Symbol for an exclusive-OR gate at A, and truth table for an exclusive-OR gate at B.

output. Any time we find all inputs of an XOR gate in the same condition (HIGH or LOW), then the output will be LOW.

SUMMARY OF GATE ACTIONS

In this chapter we have discussed five basic gates (AND, OR, NAND, NOR, and XOR), and the inverter (sometimes called a NOT gate). From these basic forms of gate, *all* digital functions can be constructed; in fact, even the largest special function LSI (large-scale integration) digital ICs contain little more than a complex interconnection of these six basic functions: AND, OR, NAND, NOR, XOR, and NOT. If we want to split hairs even further, we could say that only AND, OR, XOR and NOT are needed, because the NOR and NAND functions are NOT-OR and NOT-AND, respectively.

Figures 5-14 through 5-19 show a series of waveform timing diagrams that summarize the actions of these basic gates. Such diagrams are frequently used in digital electronics as a graphic means of showing dynamic circuit action. In fact, in many cases the timing diagram is the only reliable way to visualize circuit action.

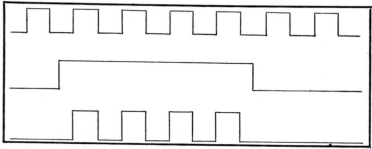

Fig. 5-14. AND gate waveforms.

In the timing diagrams, we have followed the same denotation as before: A and B are inputs, while C is the output. Note that these diagrams are for two-input devices (except, of course, for the inverter), but the basic concepts can be extended to devices with three or more inputs as well.

In Fig. 5-14, we see the operation of the AND gate. In this case, a train of square waves is applied to input A, while a single pulse is applied to input B. Note that output C remains LOW until input B goes HIGH. This is due to the requirement that both inputs be HIGH before the output can be HIGH. Once Input B has gone HIGH, because of the same rule, the output will follow input A. To emphasize again, the rules for AND gates are:

☐ The output will be LOW if either input is LOW.
☐ The output is HIGH only if both inputs are HIGH.

A timing diagram for the NAND gate is shown in Fig. 5-15. Recall that a NAND gate is an AND gate with an inverted output. The NAND gate circuit symbol shows this relationship because it is an AND gate symbol with a small circle at the output terminal.

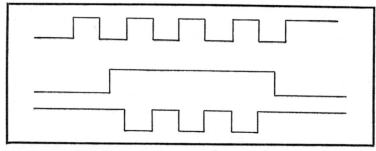

Fig. 5-15. NAND gate waveforms.

81

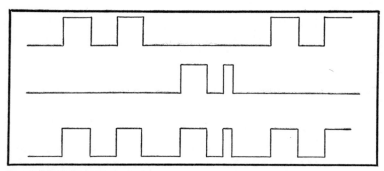

Fig. 5-16. OR gate waveforms.

The NAND gate timing diagram uses input condition similar to the previous case for the purpose of comparing their operations. A square wave is applied to input A, while a single pulse is applied to input B. If either input is LOW, then the output will be HIGH. Therefore, the output is HIGH until both A and B are HIGH. When B goes HIGH, the output will follow A, but is inverted, or 180 degrees out of phase. To reemphasize, the rules for NAND gates are:

☐ The output is HIGH if either input is LOW.
☐ The output will be LOW only if both inputs are HIGH.

The timing diagram for an OR gate is shown in Fig. 5-16. Recall that an OR gate will produce an output if either input is HIGH. This action is shown in the output waveform of Fig. 5-16. The output waveform would also be HIGH if both inputs were HIGH. To reemphasize the action of an OR gate, the rules are as follows:

☐ The output is HIGH if either input, or both inputs, are HIGH.
☐ The output is LOW only when both inputs are also LOW.

The timing diagram for the NOR gate is shown in Fig. 5-17. Recall that the output of a NOR gate will be LOW if either input is HIGH. To reemphasize, the rules for the operation of a NOR gate are:

☐ The output will be LOW if either or both inputs are HIGH.
☐ The output will be HIGH only if both inputs are LOW.

The timing diagram for an inverter is shown in Fig. 5-18. Notice that the output $\overline{A}$ is merely the inverse of the input; that is,

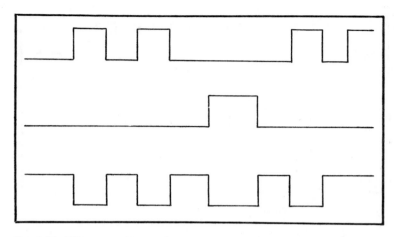

Fig. 5-17. NOR gate waveforms.

it is exactly 180 degrees out of phase with the input. Although this seems like a trivial case, it is none the less important.

The exclusive-OR gate timing diagram is shown in Fig. 5-19. Recall that the XOR gate will produce a HIGH if either input is HIGH, but not if both inputs are HIGH simultaneously. If both inputs are HIGH, or if both inputs are LOW, then the output will be LOW. This action can be seen in Fig. 5-19. Input A sees a train of square waves, and we have labeled the different sections 1, 2, 3 . . . etc. to more closely study the effects. At time 1, both inputs are LOW, so the output C is also LOW. But at time 2, input A is HIGH and input B is LOW. This will make output C HIGH until the beginning of time 3. At that time, both inputs are again LOW. Note in time period 4 that both inputs A and B are HIGH, so the output is LOW. Similarly, the output is LOW during the time period 8. To reemphasize, the rules for the exclusive-OR (XOR) gate are as follows:

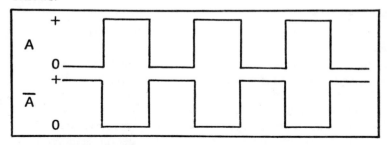

Fig. 5-18. Inverter waveforms.

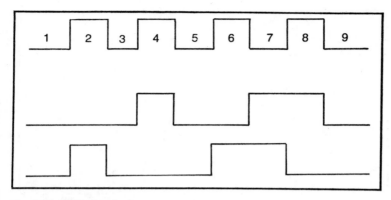

Fig. 5-19. XOR waveforms.

☐ The output is LOW any time that both inputs are in the same logic condition—HIGH or LOW.
☐ The output is HIGH if either input is HIGH, except when both inputs are simultaneously LOW.

You may have noted in some of the descriptions in this chapter that one must be careful in using the terms AND/OR and NAND/NOR. In the NAND gate description, for example, the output is HIGH if either A OR B is LOW. But isn't this NOR gate talk? No, not if we are talking about positive logic. Most of the designations are based on positive logic. But when you talk about negative logic, in which logical-1 is zero volts, and logical-Ø is V+, the definitions reverse. A positive logic NAND gate is a negative logic NOR gate. Similarly, a positive logic AND gate is a negative logic OR gate. This is why some older catalogues, published before the "standard" of positive logic was firmly established, listed the 7400 as a NAND/NOR gate; it all depends upon your point of view.

EXPERIMENTS

The experiments to follow give you the chance to examine the behavior of the various types of gate. A dual trace oscilloscope is useful in these experiments, but one can "make do" without such an instrument. You can, for example, use the logic level detector circuits shown in Appendix A, or the multichannel electronic oscilloscope switch shown in Appendix B. The oscilloscope switch will allow you to use a single trace oscilloscope to view two traces. The essential thing is not whether you know how to use an oscilloscope, but that you learn to recognize logic levels.

Experiment 5-1

A type 7404 inverter is used in this experiment. The purpose of the experiment is to show the operation of the inverter circuit—the NOT gate.

Without an oscilloscope, connect the circuit as shown in Fig. 5-20. Use the LED logic level indicators (Appendix A). These will be *on* for HIGH and *off* for LOW.

With a two-channel oscilloscope, connect the circuit in Fig. 5-21. Position the trace carrying the 7404 input on the top of the CRT screen, and the trace showing the 7404 output on the lower portion of the CRT screen.

When switch S1 is connected to the "1" position, the input of the 7404 is HIGH. When S1 is connected to the "∅" position, the input of the 7404 is grounded and is LOW.

☐ Place S1 in the "1" position and note the LEDs (or oscilloscope trace). The input is_____, so the output is_____.
☐ Place S1 in the "∅" position, and note the LEDs (or oscilloscope trace). The input is_____, so the output is_____.

Experiment 5-2

Perform the hook-up for Experiment 5-1, but replace switch S1 with the transistor switch circuit shown in Fig. 5-20. Disconnect S1 and remove it from the circuit.

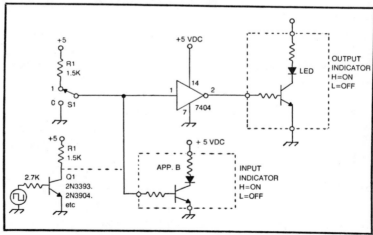

Fig. 5-20. Experiment circuit.

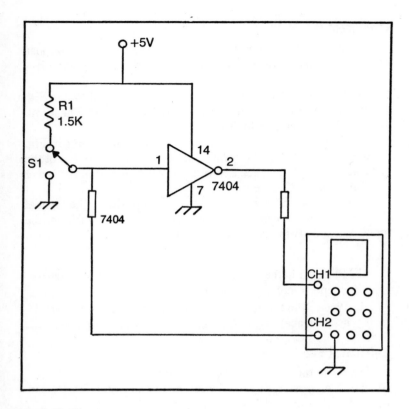

Fig. 5-21. Experiment circuit.

Connect a 1-Hz square wave generator to the input of the transistor switch circuit. Note the LEDs or the oscilloscope trace; the LEDs should be blinking back and forth as the square wave goes HIGH and LOW.

If you are using a two-channel oscilloscope, you might wish to examine the input and output waveforms at a higher frequency—over 200 hertz—and compare them with Fig. 5-18. Most oscilloscopes will be easier to trigger or synchronize at the higher frequency.

Experiment 5-3

This experiment demonstrates the 7400 NAND gate. Connect the circuit of Fig. 5-22. If you have an oscilloscope or digital logic analyzer with three or more channels, use it instead of the LED level indicators.

☐ Open both S1 and S2 (place both switches in the "1" position).
☐ The output level is (HIGH/LOW)_____.
☐ Close S1.
☐ The output is (HIGH/LOW)_____.
☐ Open switch S1 and close switch S2.
☐ The output is (HIGH/LOW)_____.
☐ Close S1 and S2.
☐ The output is (HIGH/LOW)_____.

Experiment 5-4

The 7400 TTL NAND gate IC is used to operate as an on-gate for a pulse train. The 7400 IC can be used—and often is—to control the flow of clock pulses or data into the circuit. Replace switch S2 in Fig. 5-22 with a transistor switch such as Q1 in Fig. 5-20.

Without an oscilloscope, connect a 1-Hz square wave source to the input of the transistor switch. With an oscilloscope, connect a square wave source to the input of the transistor switch. Use a frequency over 200 Hz for best oscilloscope display.

☐ Set S1 to the "0" position
☐ The output is (HIGH/LOW)_____
☐ Set S1 to the "1" position
☐ The output is (HIGH/LOW)_____

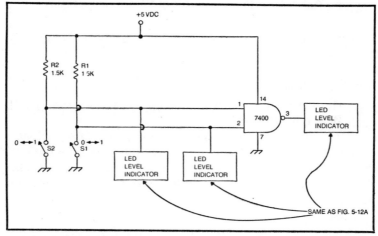

Fig. 5-22. Experiment circuit.

87

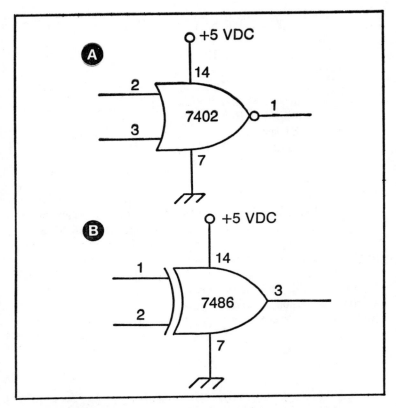

Fig. 5-23. Pinouts for 7402 at A and 7486 at B for use in experiments.

The output should pulse, or follow the input signal (inverted), *only* when S1 is set to "1."

Experiment 5-6

This experiment uses the exclusive-OR (XOR) gate. A TTL 7486 device is selected.

Perform experiments 5-3 through 5-5 using a 7486 XOR gate instead of the gates called for originally. The pinouts for the 7486 are shown in Fig. 5-23B. Compare the results in each case with the results obtained previously in each experiment and the rules for the XOR gate.

Chapter 6
Arithmetic Circuits

You're going to learn here about the half-adder, adder, and subtractor circuits. These are known as arithmetic circuits. Also, you will find out about the different types of arithmetic logic chips available to perform these functions.

SELF-EVALUATION QUESTIONS

☐ Describe the principle difference between half-adder and adder circuits.
☐ What is the function performed by the TTL 7483?
☐ What is the function performed by the TTL 7485?
☐ Draw a circuit diagram for a half-adder circuit.

ADDER CIRCUITS

There are two basic forms of adder circuit. These are called the *half-adder* and *full-adder* (or simply *adder*). Of these, several different types exist. But before examining the various types of circuits, consider the process of binary addition to find out the requirements placed on any circuit the purports to be a binary adder.

The rules for binary addition are as follows:

$$0 + 0 = 0$$
$$0 + 1 = 1$$
$$1 + 0 = 1$$
$$1 + 1 = 0 \text{ plus carry 1}$$

Examine these rules and compare them with the truth table for an exclusive-OR gate. An XOR gate obeys the following rules:

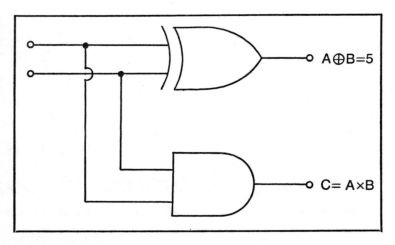

Fig. 6-1. Half-adder circuit.

☐ If both inputs are zero, then the output is zero.
☐ If both inputs are one, then the output is zero.
☐ A one output is created if either, *but not both*, inputs are one.

The truth table for the XOR gate, then, is

A	B	Output
0	0	0
0	1	1
1	0	1
1	1	0

Note that the sole difference between the truth table for the XOR gate and the rules for binary addition is that, in addition, a *carry-one* output is created if both A and B are 1. We can therefore create an addition circuit with an XOR gate and a means for generating the carry-one output.

Figure 6-1 shows a circuit for a *half-adder*. A half-adder is a circuit that generates two outputs: *sum* (S = A + B) and *carry* (C = A × B) (*read*: S = A XOR B and C = A AND B). The exclusive-OR gate generates the sum output, while an AND gate generates the carry output. In many cases, the logic symbol shown in Fig. 6-2 is used to denote the half-adder in logic diagrams, block diagrams, or flow charts.

A full adder, or simply adder as it is usually called, is a circuit that will accept as a valid input the carry output from a lower order stage (another adder). Consider again the process of binary addition. Add two binary numbers:

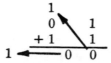

In the first step, we added the two least significant digits. In this case they are both ones, so the result is zero, and a carry output 1 is generated. This carry one is added to the next most signiificant column. In the second step, we add a 0 and a 1 to obtain a 1 result. Next we add the carry one from the previous step, so this is 1 + 1. The result (written down below the line) is a 0, with a carry-1 generated. The result of this operation is a final result of 00 and a carry 1 output. The carry 1 output can be ignored if there is no more significant digits.

The first operation in the above problem could be performed by a half-adder circuit because the least significant digit does not receive any carry bits from a lower order stage (there are no preceding stages to generate any carries). All subsequent stages, however, must be full adder circuits so that they can accommodate carries from previous stages. Otherwise, an erroneous output result could be obtained.

Figure 6-3 shows the logic block diagram for a full-adder, which by definition will:

☐ Add A and B (S = A XOR B).
☐ Account for any carry-in C from lower order stages.
☐ Produce a carry-out 1, C_o, if needed.

The circuit of Fig. 6-3 consists of two half-adders and an OR gate to generate the carry output. Half-adder HA1 adds together

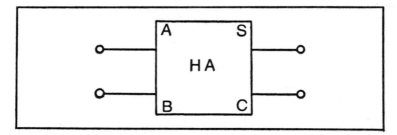

Fig. 6-2. Half-adder circuit symbol.

Table 6-1. Half-Adder Truth Table.

INPUTS		OUTPUTS	
A	B	S	C
0	0	0	0
0	1	1	0
1	0	1	0
1	1	0	1

the A and B inputs using the XOR function. This circuit will produce a carry-out 1 if needed. Half-adder HA2 adds together the results of HA1 (S′) to produce the final output: $S = S'$ XOR C_i. A carry-out may also be generated by this stage. A commonly used logic symbol for the full adder is shown in Fig. 6-4.

SUBTRACTOR CIRCUITS

Addition and subtraction are very similar arithmetic operations. In fact, it is proper to view subtraction as the addition of a positive and a negative number. It is not surprising, then, to find certain similarities between adders and subtractors used in digital circuits; in fact, the principle difference is the capability to account for a borrow-1 from the next higher order digit.

Figure 6-5 shows a simple half-subtractor circuit which produces an output difference d of A XOR B, and a borrow b of $\overline{A}$ AND B. The logic symbol for a half-subtractor is shown in Fig. 6-6.

An example of a half-subtractor circuit made entirely from standard TTL NAND, NOR, and NOT gates is shown in Fig. 6-7. This type of circuit is often found in digital equipment because the low cost of TTL IC devices makes it economical.

A full-subtractor must not only find the difference between the two inputs, and create the borrow, but must also account for

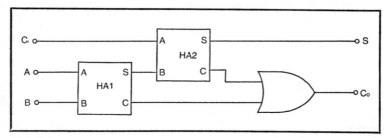

Fig. 6-3. Full-adder circuit.

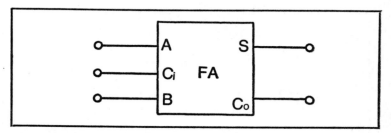

Fig. 6-4. Full-adder circuit symbol.

borrows from other stages. Fig. 6-8 shows the full-subtractor circuit, in which two half-subtractors and an OR gate are used.

A universal adder/subtractor circuit is shown in Fig. 6-9. Recall that both adders and subtractors produce an output of A

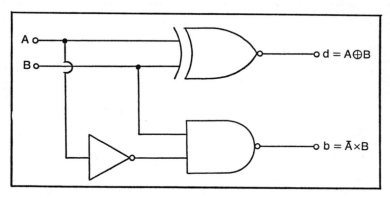

Fig. 6-5. Half-subtractor circuit.

XOR B. The principal difference is that adders produce a *carry* signal (C_o = A AND B), while subtractors produce a *borrow* signal ($b_o = \overline{A}$ AND B).

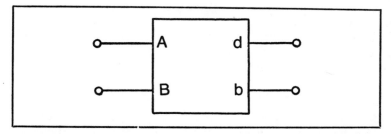

Fig. 6-6. Half-subtractor symbol.

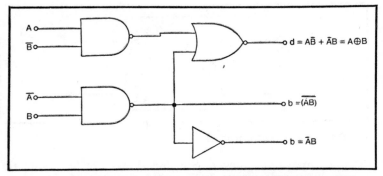

Fig. 6-7. TTL subtractor made from standard NAND, NOR and inverters.

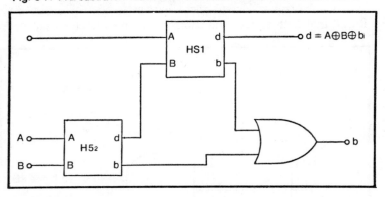

Fig. 6-8. Full-subtractor.

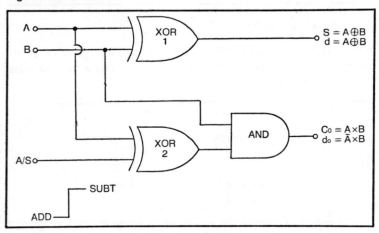

Fig. 6-9. Adder/subtractor made from XOR and AND gates.

In Fig. 6-9, the output of XOR1 will be A XOR B for both modes: addition or subtraction. The input of XOR2 marked "A/S" is the mode control terminal and selects whether the addition or subtraction is used. If A/S is LOW, then the output of XOR2 will be LOW if A is HIGH. The output of the AND gate, then, will be HIGH when A is HIGH AND B is HIGH (A XB). This is a *carry*, so the circuit operates as an adder when A/S is LOW.

When A/S is HIGH, the output of XOR2 will be HIGH when A is LOW. We can, therefore, state that the output of the NAND gate is $\overline{A} \times B$, which is the definition of a *borrow* function. The circuit of Fig. 6-9, then, operates as a subtractor when A/S is HIGH.

Chapter 7
Flip-Flops: Clocked and Unclocked

The objective of this chapter is for you to learn the operation of clocked and unclocked flip-flops. You will also become familiar with the various TTL/CMOS flip-flops available.

SELF-EVALUATION QUESTIONS

- [ ] An R-S flip-flop made with _____gates produces a Q=1 output when the S input is momentarily brought LOW.
- [ ] An R-S flip-flop made with _____gates produces a Q=1 output when the S input is made momentarily HIGH.
- [ ] A _____flip-flop transfers its input data to the output on the arrival of clock pulses.
- [ ] A J-K flip-flop goes to a state of Q=1 if the_____input is made HIGH.

All computer and digital logic functions can be made from combinations of AND, OR, NAND, NOR, and NOT gates; in fact, even the NAND and NOR functions are merely combinations of NOT gates with AND and OR gates, respectively. All of the gate functions are transient circuits; that is, they are incapable of *storing* information, even for a short period of time.

Flip-flops are circuits that will store single-bits of information. Most solid-state computer memories are arrays of flip-flops organized in a manner that allows storage of the digital words of a computer.

Most flip-flops are *bistable* circuits; that is, there are two stable output states. The flip-flop is not particularly concerned which state is in existence at any given time; it is happy in either state. If the output is labeled Q, we find that Q can be either HIGH or LOW and will be stable when it is HIGH or LOW.

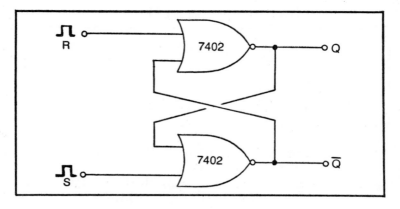

Fig. 7-1. NOR gate version of an R-S flip-flop.

R-S Flip-Flops

One of the simplest flip-flops is the *reset-set* (R-S) flip-flop. These circuits can be made from either NAND or NOR gates, although the performance characteristics are different for the two different types. The NOR gate implementation is shown in Fig. 7-1, while an occasionally used circuit symbol is shown in Fig. 7-2.

Note that the circuit has two outputs, labeled Q and $\overline{Q}$ (read "not Q"). These outputs are complementary, meaning that one will be HIGH while the other will be LOW. If Q is HIGH, then not-Q must be LOW. Similarly, when not-Q is HIGH, Q must be LOW.

In the NOR gate R-S flip-flop, the output changes state when an appropriate input is momentarily brought HIGH. Only a brief pulse at the input is needed to effect the change. The rules governing the operation of the NOR-gate R-S flip-flop are summarized in the truth table shown in Table 7-1. These rules follow.

☐ When both S and R input are LOW, no change occurs in the output.

☐ If both S and R are simultaneously brought HIGH, this is a *disallowed state*. This input condition is to be avoided.

Fig. 7-2. Circuit symbol for the R-S
flip-flop shown in Fig. 7-1.

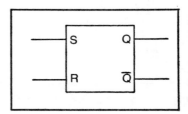

Table 7-1. Truth Table for an R-S Flip-Flop.

S	R	Q	Q̄
0	0	(no change)	
1	0	1	0
0	1	0	1
1	1	(disallowed)	

☐ Bringing the *S* input momentarily HIGH will *set* the output; i.e., make Q HIGH and not-Q LOW.

☐ Bringing the *R* input momentarily HIGH forces the flip-flop to *reset*; i.e., make Q LOW and not-Q HIGH.

The basic NOR gate R-S flip-flop can be constructed from either TTL or CMOS NOR gate ICs. In the case shown in the example, TTL type 7402 NOR gates have been used to make the flip-flop.

The NAND gate R-S flip-flop uses inverted logic; i.e., the output state changes are caused by bringing the appropriate input LOW. An example of a NAND gate R-S flip-flop is shown in Figs. 7-3 and 7-4.

Because the inputs are active-LOW, this circuit is sometimes called the $\overline{\text{reset-set}}$ (or $\overline{\text{R-S}}$) flip-flop. In this text, however, it is more convenient to simply refer to both types as R-S flip-flops, and

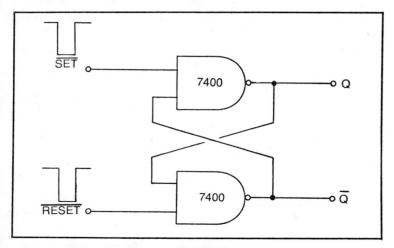

Fig. 7-3. NAND gate version of an R-S flip-flop.

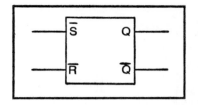

Fig. 7-4. Circuit symbol of the R-S flip-flop shown in Fig. 7-2.

then be sure that the NAND/NOR distinction is made in the text. The example of Fig. 7-3 is implemented using the type 7400 TTL NAND gate.

The output conditions are governed by the following rules and summarized in the truth table shown in Table 7-2.

☐ If the $\overline{S}$ and $\overline{R}$ inputs are both simultaneously LOW, we have a *disallowed state*. Such states must be avoided.
☐ If both S and R inputs are made simultaneously HIGH, then no change in the output state will occur.
☐ If the S input is momentarily brought LOW while the R remains HIGH, then the Q output goes to the HIGH state and not-Q is LOW.
☐ If the R input is momentarily brought LOW and the S input remains HIGH, then the Q output goes LOW and the not-Q goes HIGH.

The R-S flip-flop is used frequently in applications where a pulse to the S or R input sets a condition and then a subsequent pulse is used to reset the circuit to its initial conditions.

CLOCKED R-S FLIP-FLOPS

The operation of the two previous R-S flip-flops was unconditional. The output state changed immediately when an appropriate input pulse was received. Such circuits are only able to operate *asynchronously*.

Table 7-2. Truth Table for the NAND Gate R-S Flip-Flop.

$\overline{S}$	$\overline{R}$	Q	$\overline{Q}$
0	0	(disallowed)	
1	0	0	1
0	1	1	0
1	1	(no change)	

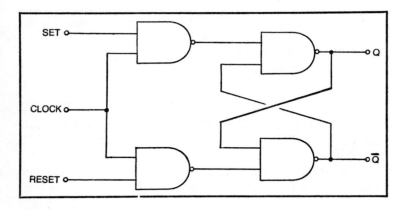

Fig. 7-5. Clocked R-S flip-flop.

The clocked R-S flip-flop of Fig. 7-5 is able to operate in a synchronous manner; i.e., the output will change state only when the input pulse coincides with a *clock* pulse. This behavior is obtained by adding a pair of NAND gates to the NAND-type R-S flip-flop circuit.

A commonly used circuit symbol for the clocked R-S flip-flop is shown in Fig. 7-6. This circuit is sometimes called an RST flip-flop.

An example of a timing diagram for an RST flip-flop is shown in Fig. 7-7. The clock produces a fixed-frequency chain of square waves at the C input. Note that the Q output does not change state immediately when the S input pulse goes HIGH. The output waits until the clock input is also HIGH. Similarly, the output is not reset until the clock pulse and the R input pulse are coincident.

MASTER-SLAVE FLIP-FLOPS

It is often difficult, or impossible, to transfer data through a circuit in an orderly manner. We often get into high-speed electronic analogies of the old-fashioned "relay race" problem. Recall that type of problem in circuits where two relays are supposed to close simultaneously? If one relay is a little sluggish or the other is

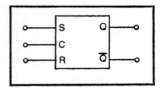

Fig. 7-6. Circuit symbol of the clocked R-S flip-flop.

100

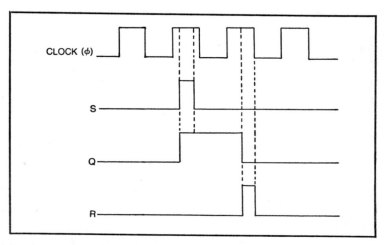

Fig. 7-7. Typical operation of a clocked R-S flip-flop.

a little faster than usual, then there will be a brief instant where one is open and the other is still closed. This condition often produces unpredictable results. The same action exists in digital circuits, and is caused by device propagation time.

Note that two RST flip-flops and an inverter can be used to synchronize a data transfer. The circuit, which is shown in Fig. 7-8, is called a *master-slave flip-flop*.

In Fig. 7-8, the inputs to FF1 are the inputs to the circuit as a whole. The outputs for the overall circuit are the outputs of FF2.

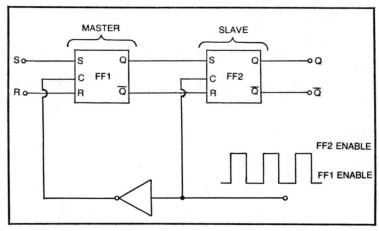

Fig. 7-8. Master-slave flip-flop.

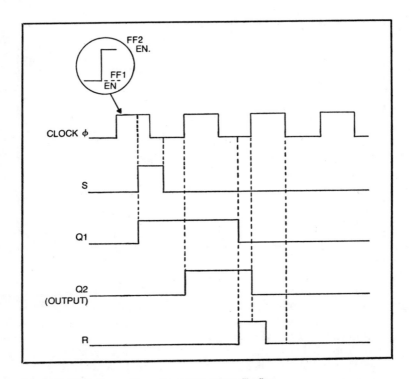

Fig. 7-9. Typical operation of a master-slave flip-flop.

Also, the clock affects FF2 directly but must pass through an inverter before it can affect FF1. Both FFs only become active when their respective clock inputs are HIGH.

A timing diagram for this circuit is shown in Fig. 7-9. Recall from above that FF1 is enabled when the clock pulse is LOW, and FF2 is enabled when the clock pulse is HIGH.

When the pulse is applied to the S input, nothing happens at Q1 (the Q output of FF1) until the clock pulse drops LOW. At that time Q1 snaps HIGH, thereby making the S input of FF2 HIGH. But at this time the clock pulse is LOW, so no change occurs at the FF2 output terminal. When the next clock pulse arrives, however, the FF2 clock input goes HIGH (Q1 is still HIGH), so the Q output of FF2 goes HIGH.

Similarly, the reset pulse arrives when the clock is LOW, so the Q1 output immediately drops LOW. The Q output of FF2, however, remains in the HIGH condition until the HIGH transition of the next clock pulse.

102

FF1 is considered the master flip-flop, while FF2 is the slave. The action at FF1 is given time to settle before the changes can be reflected at the output of FF2. This provides an orderly transfer of data between input and output.

TYPE-D FLIP-FLOPS

A type-D flip-flop is a modified RST flip-flop that has only one input (labeled the *D* or *data* input). See Fig. 7-10. The type-D flip-flop will transfer data from the D input to the Q output *only* when the clock terminal is HIGH. The following are the rules governing the operation of the type-D flip-flops:

☐ When the clock input goes HIGH, the data present on the D input is transferred to the Q output.

☐ If the clock input *remains* HIGH, the Q output will *follow* changes in the data present at the input.

☐ If the clock remains LOW, then the Q output will *retain* the data that was present on the D input at the instant the clock dropped LOW.

Because of the behavior presented in the above rules, the type-D flip-flop is sometimes called a *data latch*, or simply *latch*. The circuit symbol is shown in Fig. 7-11.

We can see these rules more graphically in Figs. 7-12 through 7-14. Consider Fig. 7-12 first. Recall that the data on the D input will be transferred to the Q output only when the clock terminal is HIGH. At time t_0 in Fig. 7-12, the clock goes HIGH, the D input is LOW, though, so the Q output is LOW also.

At time t_1, the D input goes HIGH but cannot affect the Q output because the clock is LOW. The clock goes HIGH at t_2, so the

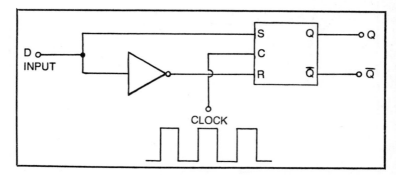

Fig. 7-10. Making a type-D flip-flop from an R-S flip-flop.

103

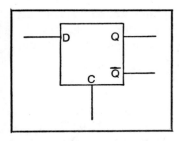

Fig. 7-11. Circuit symbol of a type-D flip-flop.

output will also go HIGH. The output pulse exists only for the interval $t_3 - t_2$, because the HIGH conditions on the D input only coincide in that interval.

At time t_5, both the D and the clock inputs go HIGH, but note that the D line remains HIGH even after the clock pulse disappears. By the third rule, then, the output must remain HIGH after the clock pulse passes.

Another situation is shown in Fig. 7-13. In this case, the clock line goes HIGH and remains HIGH. The Q output, therefore, is in an unlatched condition, so will follow the data at the D input. Because the D input data is a square wave, the output data will also be a square wave.

Still another condition is shown in Fig. 7-14. Again the D input data is a square wave, but the clock is not permanently HIGH in this example. At time t_1, both the clock and the D input are HIGH, so the Q output is also HIGH. At time t_2, the clock drops LOW, but since D is HIGH at that instant, the output will remain in the HIGH condition. Note that the D input can change at will, without affect-

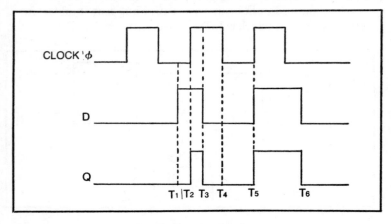

Fig. 7-12. Typical operation of a type-D flip-flop.

104

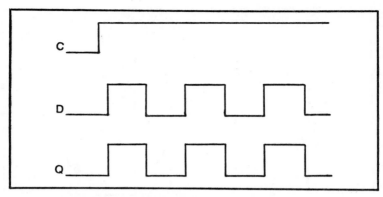

Fig. 7-13. Operation with clock HIGH.

ing the output, as long as the clock input remains LOW. But at the time t_5, the clock goes HIGH, and since the D line is LOW, the output goes LOW also. At time t_6, however, the D line goes HIGH, while the clock is still HIGH, so the output goes HIGH.

Examples of Type-D Flip-Flops

Only rarely will modern circuit designers use individual logic gates to make a flip-flop of any variety. The only common example is the R-S flip-flop. Even those, however, are commonly available in the form of standard CMOS chips, where they had not been in TTL. There are too many good integrated circuit flip-flops on the market for anyone to seriously consider building their own. Figures 7-15 through 7-18 show two TTL and one CMOS version.

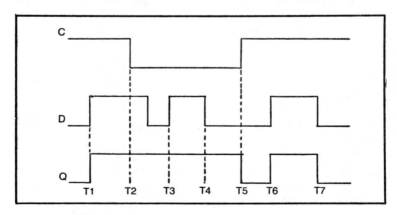

Fig. 7-14. Another operation.

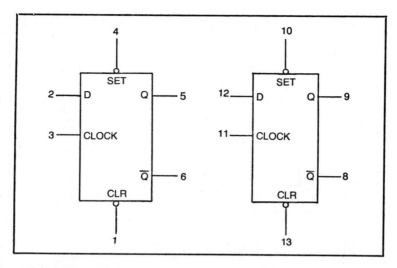

Fig. 7-15. TTL dual type-D flip-flop.

A TTL dual type-D flip-flop is shown in Fig. 7-15. The logic diagram is shown in Fig. 7-16. This particular type-D flip-flop has both *set* and *clear* inputs, in addition to the normal clock and D inputs. The 7474 device will obey the following rules of operation:

□ Data on the D input is transferred to the Q output only on positive-going transitions of the clock pulse.

□ No changes in the data are reflected as output changes unless the clock is *going* HIGH. The device is, therefore, an edge-triggered flip-flop.

□ If the *set* input is grounded, then Q immediately goes HIGH and not-Q goes LOW.

□ If the *clear* input is grounded, then Q goes LOW and not-Q goes HIGH.

□ Set and clear inputs must not be simultaneously grounded. If these inputs are not being used, then they should be tied HIGH to +5 VDC.

□ The two-flip-flops in the 7474 IC are completely independent of each other, except for sharing common power supply and ground terminals.

The TTL 7475 device shown in Fig. 7-17 is a special case of the type-D flip-flop called a *quad-latch*. This IC device is level sensitive, so the output will follow changes as long as the clock terminal is HIGH. This is more like the behavior expected of traditional type-D flip-flops, as given earlier in this section.

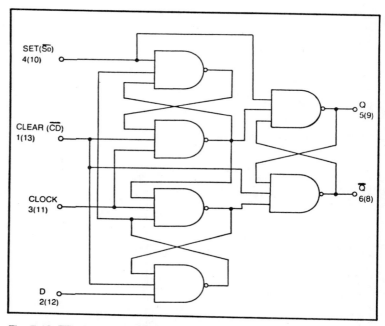

Fig. 7-16. TTL dual type-D flip-flop.

The 7475 is used primarily to hold four bits of data; i.e., one bit in each flip-flop. The pinouts for the 7475 are shown in Fig. 7-17. The clock inputs are labeled *enable* inputs. When the enable inputs are HIGH, then, data on the D inputs are transferred to the Q outputs. The 7475 is arranged in a 2 × 2 format, meaning that the enable inputs of two flip-flops are tied together and are brought out to a package pin. The enable inputs to the other two remaining

Fig. 7-17. TTL quad latch is a kind of type-D flip-flop.

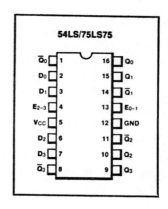

107

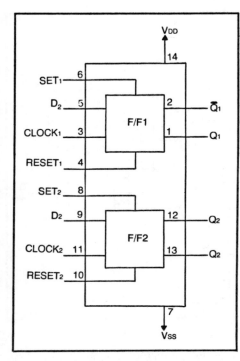

Fig. 7-18. CMOS dual type-D flip-flop.

flip-flops are also tied together and are brought out to another package terminal.

One common place to find 7475s is in digital counter applications, where they are used to hold the BCD data being displayed while the counter chip, which is often a 7490 device, updates the data. The display can be updated when the count is completed by momentarily bringing the enable terminals HIGH. A single 7475 will store all four bits required for a single BCD display decoder.

A device similar to the 7475 is the TTL IC type 74100. This 24-pin IC contains a pair of four-bit latches and therefore has the ability to store up to eight bits of digital data, a fact that is not lost on microcomputer circuit designers.

A CMOS type-D flip-flop is the CD4013 or, simply 4013, device shown in Fig. 7-18. Like the 7474 device from the TTL line, the CMOS CD4013 is a dual type-D flip-flop. It can be used in either direct or clock modes, i.e., it has set and clear direct inputs as well as D and clock inputs. The 4013 differs from the 7474, however, in that direct inputs become active when HIGH. If usused, therefore, these inputs should be tied to *ground*, instead of

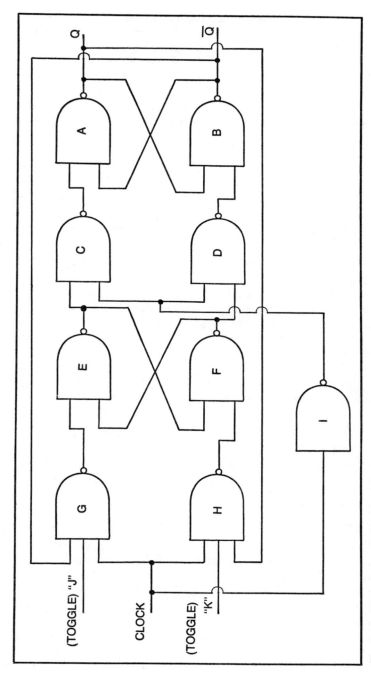

Fig. 7-19. J-K flip-flop made from NAND gates.

Table 7-3. J-K Flip-Flop Truth Table.

INITIAL CONDITIONS				FINAL CONDITIONS	
OUT		IN			
Q	$\overline{Q}$	J	K	Q	$\overline{Q}$
0	1	0	0	0	1
0	1	0	1	0	1
0	1	1	0	1	0
0	1	1	1	1	0
1	0	0	0	1	0
1	0	0	1	0	1
1	0	1	0	1	0
1	0	1	1	0	1

+5 volts, as in the case of the TTL version. This is exactly the opposite protocol from the TTL.

In the clocked mode, the 4013 behaves much like the 7474. It is a positive-edge triggered device, so data is transferred to the Q output when the clock line goes HIGH. The action occurs on the positive-going *transition* of the clock pulse. In this respect, the 4013 and 7474 follow the same rules.

In the direct mode, the operation is similar to the 7474 but requires pulses of the opposite polarity. The rules for the set and clear inputs of the 4013 are as follows:

☐ If the set input is made HIGH, then the Q output goes HIGH and not-Q goes LOW.

☐ If the clear input is made HIGH, then the Q output goes LOW and not-Q goes HIGH.

J-K FLIP-FLOPS

The J-K flip-flop is very similar to the other clocked flip-flops, although certain differences exist that make the J-K special. Figure 7-19 shows a gate logic diagram for a J-K flip-flop, while the usual schematic symbol is shown in Fig. 7-20. Table 7-3 shows the truth table. Like the TTL and CMOS type-D flip-flops discussed earlier, the J-K flip-flop is capable of operating in either direct or clocked modes.

Direct operation uses the set and clear inputs to force the Q and not-Q outputs into specific states—HIGH or LOW. The truth

110

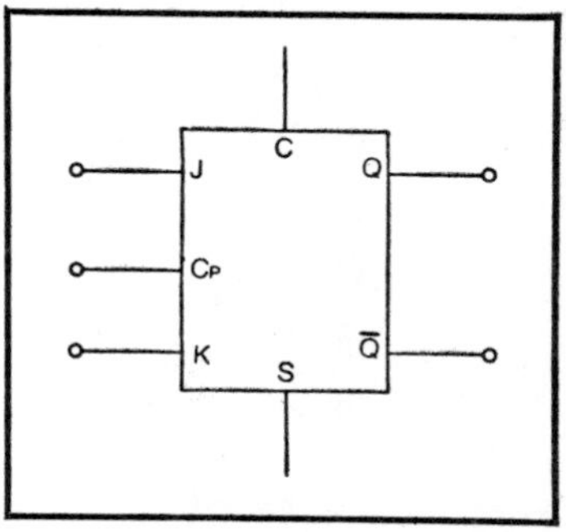

Fig. 7-20. Circuit symbol for a J-K flip-flop.

table for TTL J-K flip-flops (7473, 7476, etc.) is shown in Table 7-4. The rules for these devices are:

☐ LOW conditions on both set and clear inputs results in a forbidden, or disallowed, state.
☐ Making the clear input HIGH and the set input LOW forces Q HIGH.
☐ Making the clear input LOW and the set input HIGH forces Q LOW.
☐ Making both set and clear inputs HIGH causes the flip-flop to operate in the clocked mode.

Again the CMOS version operates in a similar manner, but with opposite polarity pulese. The CD4027 is a dual J-K flip-flop, so:

☐ If both set and clear are LOW, then normal clocked operation is obtained.
☐ Making clear LOW, and set HIGH, forces the Q output HIGH.
☐ Making clear HIGH, and set LOW, forces Q LOW.
☐ Making both set and clear HIGH is a disallowed condition.

Table 7-4. Truth Table for Unclocked Operation.

SET	CLEAR	Q
0	0	(disallowed)
0	1	1
1	0	0
1	1	(normal condition for clocked operation)

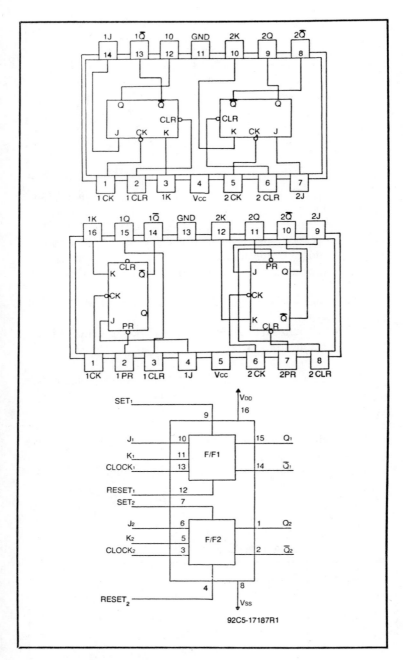

Fig. 7-21. IC flip-flops.

Table 7-5. Truth Table for Clocked Operation.

SET	CLEAR	CLOCK	Q
0 0 1 1	0 1 0 1	⌐▼_	No change 0 1 Goes to opposite

Clocked operation means that all changes occur synchronously with the clock pulse. The J-K flip-flop operates on the negative going transition of the clock pulse, in distinct contrast to the behavior of he 7474 (type-D) device discussed earlier.

Table 7-5 shows the operation of TTL J-K flip-flops in the clocked mode. The changes that are shown occur as illustrated on the negative transition of the clock pulse. The rules of operation are as follows:

☐ If both both J and K are LOW, no change occurs in the output, regardless of clock pulse transitions.
☐ If J is LOW and K is HIGH, the Q output is forced LOW.
☐ If J is HIGH and K is LOW, Q is forced HIGH.
☐ If both J and K are HIGH, the Q output will go to the *opposite* of its present state; for example, if it is HIGH, then it will go LOW, and vice versa.

In the clocked mode, the CMOS 4027 device follows these same rules, except that the changes occur on the positive-going transition of the clock pulse.

J-K Flip-Flop Examples

We have mentioned three examples of commercial IC J-K flip-flops have been mentioned: 7473 (TTL), 7476 (TTL), and 4027 (CMOS). Because the rules for these devices have already been given, they will not be repeated here. Figure 7-21 shows the pinouts for the 7473, 7476, and 4027 devices, respectively.

Chapter 8
Some Advice

Technicians servicing digital equipment very often find themselves trying to figure out how the circuit works. Unfortunately, much equipment is on the market with little or no documentation. In one case, a supplier of a major medical computer system sold a complicated installation and then could not supply either service or *operator's* instruction manuals! Some manuals are little better than no manual. We have all seen those situations where we are at first elated to see a genuine service manual, only to find out a short time later that the service manual is darn near useless. Some manufacturers seem to have a rather bad attitude towards servicers, especially independents. Their attitude is reflected by the nature of their manuals and other "field support." Note that most of these firms are not exactly hostile towards the field service industry, but simply don't care, unless their customers throw a fit over service problems. If you are a business person seeking to establish relationships with digital equipment manufacturers that will allow you to service their equipment, ask first to see the existing documentation. Find out whether or not you can live with their idea of "field support." This might well affect your contract negotiations. If the circuit level documentation is poor or scanty, for example, you might want to be responsible only for board swapping in the field, leaving all of the actual component level troubleshooting to factory hands. But if they are able to provide a lot of detailed information about their product, in the form of schematics, PC board layouts and circuit operation descriptions, then consider offering them service to the component level.

POOR DOCUMENTATION

But right now, the principal concern is to try to figure out what to do in those cases when the documentation is too poor for usefulness or is nonexistent, and the job *has* to be done regardless. That's not an easy order to fill, and nobody has any panacea.

The first advice is to ask you to learn as much as you can about the individual circuit elements; i.e., logic gates, flip-flops, etc. Buy all of the standard IC device reference manuals on digital chips. This publisher will be happy to send you their latest catalog of books. Also buy all of the manufacturer's data books on logic ICs. Motorola, Texas Instruments, National Semiconductor, Signetics, RCA, Intersil, and Fairchild Semiconductor all publish comprehensive data books. A lot of duplication is among them, but they are all worthwhile in their own right. The reason is that *no one* makes completely universal lines of ICs. Each might have a few holes not covered by a data book. Also, each might have a few special "house number" devices not covered in other manufacturers' data books. Your library could save you an awful lot of trouble. A $100 investment in books every couple of years is not too great to ask of a business. After all, it will yield much in the way of time savings and completed jobs.

Another good piece of advice is to learn *well* the operation of circuits. For this, you might want to start with this basic textbook. Read as many books and magazine articles as time permits. Read them even if you are not particularly interested in the topic of the article itself. All circuits are constructed of building blocks. The device itself may be of no interest to you at the time, but the author might describe circuit techniques and building blocks that *are* of interest.

You might have guessed that I am a fan of reading and recommend it highly for everybody. I have noticed over the past two decades that the best service technicians are those who read hobby and trade magazines and books. This might seem a little bit self-serving, coming from an author, but it is true. I invite you to look around at the successful techs that you know.

In almost all cases, even in asynchronous circuits, the key to figuring out what is happening is *timing*. In all digital circuits, timing is critical. If the circuit is too complex for you to analyze in your head, then draw a timing diagram of how it should work. You have seen these diagrams all through this book. Use the rules of operation for the individual devices and make a diagram of the expected pulses and logic levels. Confirm them on an oscilloscope. This procedure does take time but is sometimes your only hope. Enshrine this concept deeply in your mind:

TIMING IS THE KEY!

Certain relationships exist between the different binary logic functions (OR, AND, NAND, NOR, XOR, NOT). These are described mathematically using logical notation, something called Boolean algebra and a tool called deMorgan's theorem. I do not believe that these matters are necessary in the training of good digital service technicians, but they are useful, if only for the purpose of helping you gain insight into how some circuits operate. We are not going to discuss the mathematical aspects here but do recommend that you consider reading a book that does cover such matters.

Do not be afraid to ask manufacturer's service personnel for advice on certain problems. They usually know the equipment better than anyone else. Some companies, though, seem to have a standing policy of referring all incoming telephone calls to the *sales department*! They are useless to you in most cases. Customer service offices are sales departments in disguise that have nothing at all to do with the kind of service department that you need. The exception is when you need to manipulate their built-in clout to gain access to a technician. Try to bypass sales and management (even the service manager) and get on the phone with someone who repairs the equipment—a real, live hands-on person. Service department personnel are usually less inhibited about talking about their equipment problems than are sales and management people.

Incidentally, the engineering department is occasionally useful to field servicers, especially in solving really exotic problems. Most of the time, however, the engineering department is of little use to the servicer. It's not that they are stupid, of course, or too theoretical; they are too far removed from normal day-to-day service problems. There is also a built-in bias on the part of in-house engineering departments that everybody in the field is stupid. That's another problem you don't need. Besides, most engineers are too busy with *future* products to be much concerned over current or past products.

SOMETHING FOR TV SHOPS

The modern solid-state color TV receiver has caused some changes in the service industry. Besides some color TV sets now containing substantial amounts of previously unfamiliar digital electronics circuits, especially in tuning, remote control, and on-screen data display, there are certain other aspects to the change. One is that the complexity of the sets has increased substantially over the years. This situation places increasing demands on the service personnel. No longer can a "tube jockey" repair substantial numbers of the shop's total work load. The same level of knowledge is now required for a TV servicer as was formerly required only in commercial, industrial, and military electronics work. Yet at the same time, TV set prices have not increased appreciably over the years. The real price, measured in how long someone has to work to buy the product, is therefore *reduced* because of inflation. To the servicer, these factors mean that a more expensive technician is required to repair a less expensive product. This usually makes the customer less inclined to pay a reasonable price for service.

One answer to this dilemma is for the service shop to take its highly qualified people and go into some other branch of service work. The explosive growth of digital equipment and its seemingly ubiquitous distribution makes digital electronics a natural consideration. You will, find the commercial, industrial, and institutional (i.e. hospitals, governments, etc.) more able and willing to pay for good quality service. As I am writing this chapter, the going rate for a TV service call in the Washington, DC area is around $25, and that is collected only with some trepidation. At the same time, commercial service rates are at a $35 to $40 per hour level. The difference becomes even more pronounced when you consider that the $25 collected by the TV servicer is a flat rate, one-time charge for the entire call. The commercial servicer operates on a portal-to-portal basis, and *he* gets to say how far away the portal was! This means that on an average two-hour service call the commercial tech collects as much as $80, while the TV tech with about the same level of expertise collects only $25. There is little wonder why commercial work pays both the company and the technician more money.

Commercial electronics servicing has both advantages and disadvantages. One advantage is that your shop facilities may be located in lower rent warehouse sites or industrial park sites, rather than in high-priced suburban shopping malls. I know one company that operates out of the basement loading dock/garage area of a shopping center. The rent is less than *one-fifth* of the rent of the same amount of floor space upstairs in the mall. Not all are that good, however! You have to have a decent front to build a good, positive image in the mind of the consumer. Most commercial service contacts are by telephone or in person. There is practically no walk-in trade at all.

But that in-person bit means that your image problem is

merely transferred from the physical plant to the technicians. In consumer service work, the techs can get away with open-collar sports shirts and dungarees. Commercial work, however, often demands a snappy, professional look. It has been claimed that magazine publisher and foreseer of much of today's electronics gadgetry, Hugo Gernsback, coined the term *servicer* to replace the *repairman* title used by radio technicians. Now there is ample justification to once again change titles—this time to *field engineer, service engineer,* or *electronic* technologist, which is the one I personally like. (In some states, the word *engineer* is reserved only for professional engineers licensed to practice engineering).

The importance of names, titles, and the personal appearance of the technicians is more than a little ego trip. It is critical to the way you are regarded by the client. They are less likely to pay higher prices for the services of someone who looks like a plumber, unless their need for the service is of the same urgency as for a plumber. In this market, professionalism is required, and one must look professional. Consider, for example, some of service departments of the big companies, such as IBM, Xerox, Digital Equipment Corporation, etc. I typed this section on an IBM *Selectric II* typewriter. In the year or so that I have owned it, IBM has serviced it three times. All of the persons ("persons" because one was a woman) who came to my house to repair, adjust or install my Selectric II have been dressed in *business suits*. And typewriter repair can be *dirty* work! When it came time to handle something messy, each would put on an apron; however, they always reported in proper business attire. They even carried their tools in a briefcase, rather than an ordinary tool box.

Chapter 9
Digital Counters:
Devices and Circuits

A digital counter is a device or circuit that operates as a frequency divider. The most basic counter is the simple J-K flip-flop connected such that J and K are tied HIGH. This makes the output produce *one* output pulse for every *two* input pulses. It is therefore a *binary*, or *divide-by-two, counter*. We can generalize, saying that a digital counter is a circuit that produces a single output pulse for x number of input pulses.

You are probably familiar with digital frequency counters, which are test instruments normally used to measure frequencies from radio transmitters. These instruments contain *decade*, or divide-by-10, counters. The generic term "counter," however, applies to a wide range of circuits that fit the definition in the previous paragraph.

There are two basic classes of digital counter circuits, *serial* and *parallel*. The serial types are all examples of *ripple* counters, which means an input change must ripple through all stages of the counter to its proper point. Parallel counters are also called *synchronous counters*. Ripple counters are operated serially, which means that the output of one stage becomes the input for the following stage.

The basic element used in counters is the J-K flip-flop, which is shown in Fig. 9-1. Note in the figure that the J and K inputs are tied HIGH and therefore remain active.

A timing diagram in Fig. 9-2 shows the action of this circuit. When a J-K flip-flop is connected as in Fig. 9-1, its output changes state on the negative-going transition of the clock pulse. In Fig. 9-2, the first negative-going clock pulse causes the Q output to go HIGH. It will remain HIGH until the input sees another negative-going clock pulse, at which time the output will drop LOW again.

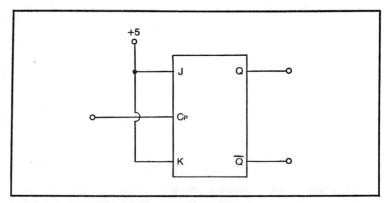

Fig. 9-1. J-K flip-flop wired for clocked operation.

The actions required to complete the output pulse take two input clock pulses. This J-K flip-flop, therefore, divides the clock frequency by two, making it a *binary* counter.

A binary ripple-carry counter can be made by cascading two or more J-K flip-flops, as shown in Fig. 9-3. This particular circuit uses four J-K flip-flops in cascade. Any number, however, could be used. A problem with this simple type of counter is that only division ratios that are powers of two (2^n) can be accommodated. In all cases, the division ratio will be 2^n, where n is the number of flip-flops in cascade. In this circuit, therefore, there is a maximum division ratio of 2^4, or 16. Of course, if we can get to the Q outputs of all four J-K flip-flops, we can create frequency division ratios of 2, 4, 8 or 16.

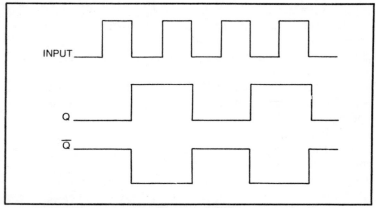

Fig. 9-2. Operating waveforms. Note the F/2 output.

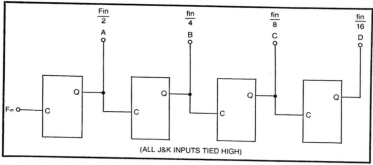

Fig. 9-3. Cascaded J-K flip-flops form a counter.

Frequency division is one use for such a counter. You may, for example, want to build a *prescaler* or some other type of divider. You can use a circuit such as shown in Fig. 9-3 for any division ratio that is a power of two (2, 4, 8, 16, 32 . . . 2^n).

But prescaling and most other simple frequency division jobs are but one example of counter applications. We can also use the circuit to store the total number of input pulses; the job most people mean when they think of "counters." Consider again the circuit of Fig. 9-3 and the timing diagram of Fig. 9-4. Outputs A, B, C, and D are coded in *binary*, with A being the least significant bit and D being the most significant bit. These outputs are weighted in the following manner: $A = 2^\phi$, $B = 2^1$, $C = 2^2$, $D = 2^3$. This is called the 1-2-4-8 system because these figures evaluate to $A = 1$, $B = 2$, $C = 4$, and $D = 8$. Recall from our discussion in Chapters 2 and 3 that these are the weights assigned to the binary number system,

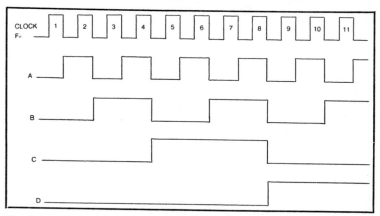

Fig. 9-4. Timing diagram of cascaded J-K flip-flops.

121

Table 9-1. Binary Code.

After Pulse No.	Word
0	0000
1	0001
2	0010
3	0011
4	0100
5	0101
6	0110
7	0111
8	1000
9	1001
10	1010
11	1011
12	1100
13	1101
14	1110
15	1111

By arranging the digits in the form *DCBA*, a binary number is created that denotes the number of pulses that have the input.

Consider the timing diagram of Fig. 9-4. Note that all Q output changes occur following the arrival of each pulse. After pulse No. 1 has passed, the Q_A line is HIGH and all others are LOW. This means that the binary word on the output lines is 0001_2 (1_{10}); *one* pulse is passed.

Following pulse No. 2, we would expect 0010_2 (2_{10}) because *two* pulses have passed. Note that Q_B is now HIGH and that all others are LOW. The digital word is, indeed, 0010_2. If you follow each pulse, you will find the binary code to be as shown in Table 9-1.

The counter shown in Fig. 9-3 could be called a *modulo-16 counter*, a *base-16 counter*, or a *hexadecimal counter*. All of these terms mean substantially the same thing.

The output of a hexadecimal counter can be decoded to drive a display device that indicates the digits Ø through 9, A through F—the hexadecimal digits. In most applications where someone is to read the output, however, a *decimal* counter is needed. We human beings use the decimal number system because of our 10 fingers.

DECIMAL COUNTERS

A decimal counter operates in the base-10, or *decimal*, number system. The most significant bit of decimal counter produces one output pulse for every 10 input pulses. Decimal counters are also sometimes called *decade counters*. The decimal counter forms the basis for digital event, period and frequency counter instruments.

The hexadecimal counter shown in Fig. 9-3 is not suitable to decimal counting unless modified by adding a single 7400 NAND gate. Recall that a TTL J-K flip-flop uses inverted inputs for the *set* and *clear* functions. As long as the *clear* input remains HIGH, the flip-flop will function normally. When the *clear* input is momentarily brought LOW, the Q output of the flip-flop goes LOW.

The decade counter shown in Fig. 9-5 is connected so that all four *clear* inputs are tied together to form a common clear line. This line is connected to the output of a TTL NAND gate, which is one section of a 7400 IC device. Recall the rules of operation for the TTL NAND gate: If either input is LOW, then the output goes HIGH, but if both inputs are HIGH, then the output goes LOW.

The idea behind the circuit of Fig. 9-5 is to clear the counter to ØØØØ following the tenth input pulse. Let's examine the timing diagram of Fig. 9-6 to see if the circuit performs the correct action. Up until the tenth pulse, this diagram is the same as for the base-16, or hexadecimal, counter discussed previously.

The output of the NAND gate will keep the *clear* line HIGH for all counts through 10. The inputs of this gate are connected to the B and D lines. The D line stays LOW, which forces *clear* to stay HIGH, up until the eighth input pulse has passed. At that time (T_o in Fig. 9-6), D will go HIGH, but B drops LOW. We still have at least one input of the NAND gate (line B) LOW, so the *clear* line remains HIGH.

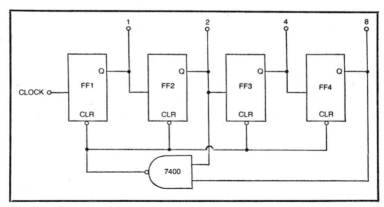

Fig. 9-5. Decade counter.

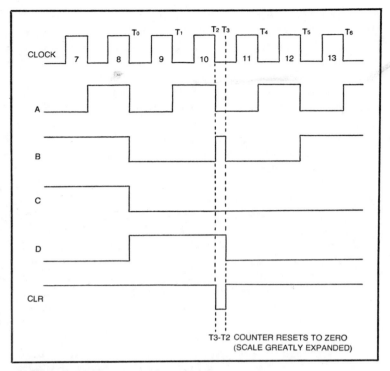

T3-T2 COUNTER RESETS TO ZERO
(SCALE GREATLY EXPANDED)

Fig. 9-6. Timing diagram a decade counter.

The *clear* line remains HIGH until the end of the tenth input pulse. At that point (T_2), both B and D are HIGH, so the NAND gate output goes LOW, clearing all four flip-flops. This forces them to go to the state where all Q outputs are LOW. The counter is now reset to 0000.

The reset counter produces a 0000 output, so both B and D are now LOW, and this forces the *clear* line HIGH once again. The entire reset cycle occurs in the period (T_3-T_2). This period has been expanded greatly in Fig. 9-6 for simplification. Actually, the reset cycle takes place in nanoseconds, or at the most, microseconds—whatever time is required for the slowest flip-flop to respond to the clear command.

The eleventh pulse will increment the counter one time, so the output will indicate 1 (0001). This counter, then, will count in the sequence 0-1-2-3-4-5-6-7-8-9-0-1-. . . . The output code is a 10-digit version of four-bit binary, or hexadecimal. It is called *binary coded decimal* (BCD).

SYNCHRONOUS COUNTERS

Ripple counters suffer from one major problem: speed. The counter elements are wired in cascade, so an input pulse must ripple through the entire chain before it affects the output. A synchronous counter feeds the clock inputs of all flip-flops in parallel. This results in a much faster circuit.

Figure 9-7 shows the partial schematic for a synchronous binary counter. Synchronous operation is accomplished by using four flip-flops, with their clock inputs tied together, and a pair of AND gates.

One AND gate is connected so that both Q1 and Q2 are HIGH before FF3 is active. Similarly, Q2 and Q3 must be HIGH before FF4 is made active. On a clock pulse, any of the four flip-flops scheduled to change will do so *simultaneously*. Synchronous counters attain faster speeds, although ripple counters seem to predominate the common counter applications.

PRESET COUNTERS

A preset counter increments from a preset point other than 0000. For example, suppose we wanted to count from 5_{10} (0101_2). We could preset the counter to 0101_2 and increment from there. The counter output pattern will be:

5	0101	9	1001	3	0011
6	0110	0	0000	4	0100
7	0111	1	0001	5	0101
8	1000	2	0010	.	.

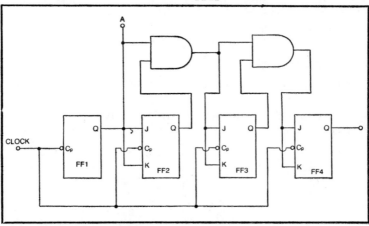

Fig. 9-7. Synchronous binary counter.

125

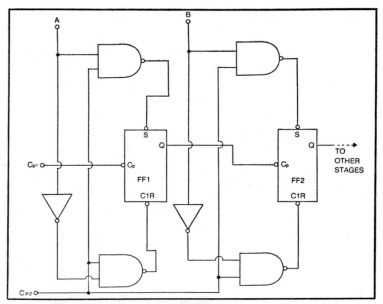

Fig. 9-8. Preset inputs using jam technique.

Figure 9-8 shows a common method for achieving preset conditions: the *jam input*. Only two stages are shown here for the sake of simplicity, but adding two additional stages will make it a four-bit counter. Of course, any number of stages may be cascaded to form an N-bit preset counter.

In Figure 9-8, the preset count is applied to A and B, and both bits will be entered simultaneously when clock line *CP2* is brought HIGH. Line *CP2* is sometimes called the *enter* or *jam* input. Once the preset bit pattern is entered, the counter will increment from these with transitions of clock line *CP1*.

DOWN AND UP-DOWN COUNTERS

A *down counter* decrements, instead of increments, the count for each excursion of the input pulse. If the reset condition is 0000, then the next count will be 0000 − 1, or (1111). It would have been 0001 in an ordinary up-counter. The count sequence for a four-bit down counter is:

0	0000	11	1011	6	0110	1	0001
15	1111	10	1010	5	0101	0	0000
14	1110	9	1001	4	0100	.	.
13	1101	8	1000	3	0011	.	.
12	1100	7	0111	2	0010	.	.

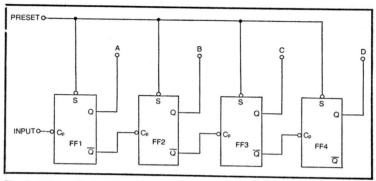

Fig. 9-9. Four-bit down binary counter.

Basically the same circuit is used as before, but toggle each flip-flop from the not-Q of the preceding flip-flop. An example of a four-bit binary down counter is shown in Fig. 9-9. Note that the outputs are taken from the Q outputs of the flip-flops, but toggling is from the not-Q.

The *preset* inputs of the flip-flops are connected together to provide a means to preset the counter to its initial (1111) state. This counter is also called a *subtraction counter*, because each input pulse causes the output to decrement by one bit.

A *decade* version of this circuit is shown in Fig. 9-10. As in the case of the regular decade counter (the up-counter), a NAND gate is added to the circuit to reset the counter following the tenth count. The states are detected when C and D are HIGH, and the two middle flip-flops are cleared. This action forces the output to

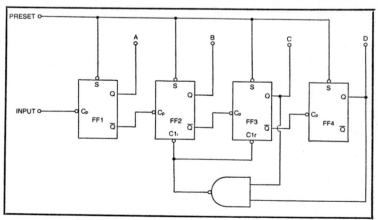

Fig. 9-10. Decade down counter.

127

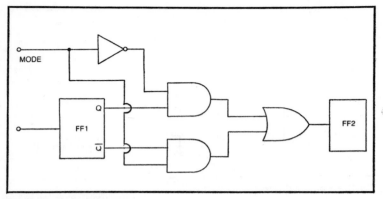

Fig. 9-11. Up/down counters.

1001_2 (9_{10}). The counter then decrements from 1001 in the sequence:

9	1001	5	0101	1	0001
8	1000	4	0100	0	0000
7	0111	3	0011	9	1001
6	0110	2	0010	.	.

UP/DOWN COUNTERS

Some counters will operate in both up and down modes, depending upon the logic level applied to a *mode* input. Figure 9-11 shows a representative circuit, in which the first two stages of a cascade counter are modified by the addition of several gates. If the mode input is HIGH, then the circuit is an UP counter. If the mode input is LOW, however, then the circuit is a DOWN counter.

TTL/CMOS EXAMPLES

Very few digital circuit designers construct counters from individual flip-flops. Too many ready-built IC counters are available in all of the major IC logic families.

There are three basic counters commonly available in the transistor-transistor logic (TTL) line: 7490, 7492 and 7493. The 7490 is a decade counter, the 7492 is a divide-by-12 (also called modulo-12) counter, and the 7493 is a base-16 (hexadecimal) binary counter. All three of these counters are of similar construction, and their respective pinouts are shown in Fig. 9-12.

The 7490 is a *biquinary* type of decade counter. This means that it contains a single, independent, divide-by-two stage. This is

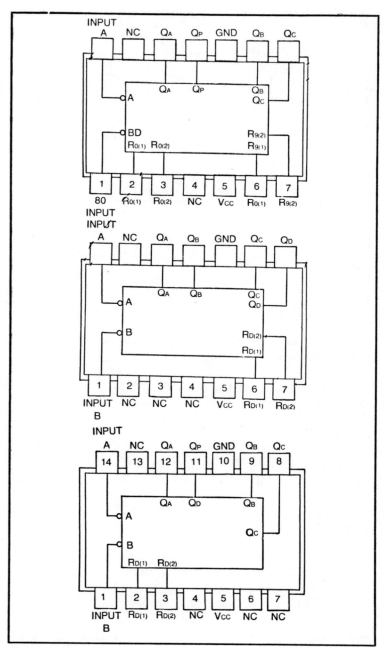

Fig. 9-12. TTL IC counters.

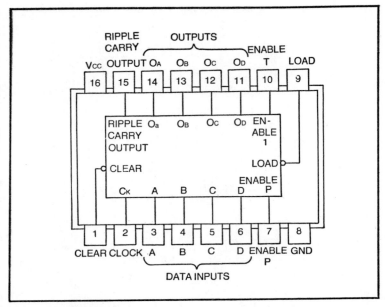

Fig. 9-13. 74160 to 163 counter.

followed by an independent divide-by-five stage. Decade division is accomplished by cascading the two stages.

Both 7492 and 7493 follow similar layout schemes. In both, the first stage is a single divide-by-two flip-flop, followed by divide-by-six (7492) or divide-by-eight (7493) stages. These form divide-by-12 and divide-by-16 counters, respectively.

The 74142 is a special function TTL IC that contains a divide-by-10 counter (BCD), a four-bit latch circuit, and a display decoder suitable for driving a *Nixie®* (registered trademark of the Burroughs Corporation). The 74142 is housed in a standard 16-pin DIP package.

Figure 9-13 shows the pinouts for the TTL type 74160 through 74163 devices. These are BCD and binary four-bit synchronous counters:

□ 74160—Decade (BCD) synchronous, direct-clear.

□ 74161—Binary, synchronous, direct-clear.

□ 74162—Decade (BCD), fully synchronous.

□ 74163—Binary, fully synchronous.

These counters typically operate to 32 MHz and dissipate approximately 325 mW. All are housed in standard 16-pin DIP packages.

These four counters are different from those that have been discussed previously, because they are divide-by-N counters, where N is an integer. The value N is applied to the *data inputs* and loaded into the counter when the *load* terminal is momentarily brought LOW.

Examples of basic CMOS counters are shown in Fig. 9-14. Again, these examples are not exhaustive, but merely representative of those commonly used in electronic circuits. Not one of them is a multidigit decimal counter such as the eight-digit, 10-MHz device made by Intersil, Inc.

□ 4017. This device is a fully synchronous decade counter, but the outputs are decoded 1-of-10. The active output is HIGH while the inactive outputs are LOW. The 4017 is positive-edge triggered. The *reset* and *enable* inputs are normally held LOW. If the *reset* is momentarily brought HIGH, the counter goes immediately to the zero state. The *enable* input is used to inhibit the count without resetting the device; that is, if the *enable* input is made HIGH, the count ceases and the output remains in its present state. The output terminal produces a pulse train of $F_{in}/10$, which is HIGH for counts 0, 1, 2, 3, and 4, and LOW for 5, 6, 7, 8, and 9.

□ 4018. This device is a synchronous divide-by-N counter, where N is an integer of the set 2, 3, 4, 5, 6, 7, 8, 9, and 10. It is difficult to decode the outputs of this counter, so its principal use is in frequency division. For normal running, the *reset* and *load* inputs must be held LOW. The 4018 is positive-edge triggered. The N-code for determining the division ratio is set by connecting the input terminal to an appropriate *output* or in certain cases, an external AND gate. For *even* division ratios, no external gate is needed; merely connect the *input* terminal as shown in Table 9-2.

The odd division ratios, such as 3, 5, 7, and 9, require an external, two-input AND gate. The 4018 outputs are connected to the AND gate inputs, and the AND gate output is connected to the 4018 input. See Table 9-3.

Table 9-2. Input Terminal Connections to the 4018 Counter.

N	Connect Input (Pin No. 1) To	Pin No.
2	Q1	5
4	Q2	4
6	Q3	6
8	Q4	11
10	Q5	13

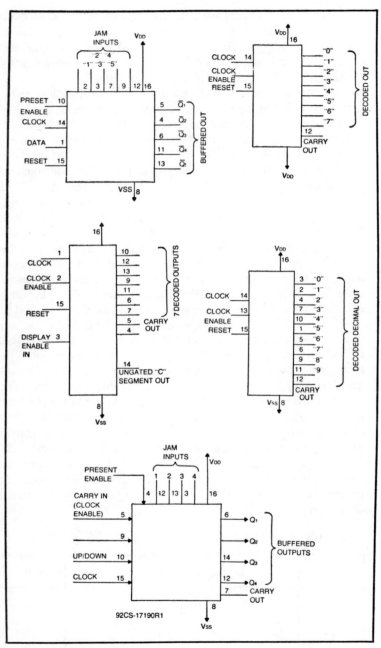

Fig. 9-14. CMOS counters.

Table 9-3. Output Terminal Connections to the 4018 Counter.

N	Connect To AND Gate Inputs	Pin Nos.
3	Q1, Q2	5, 4
5	Q2, Q3	4, 6
7	Q3, Q4	6, 11
9	Q4, Q5	11, 13

The feedback line just described is also the main output from the counter. If, for example, "input" pin No. 1 will be (by the table above) 1/6 of the clock frequency.

The 4018 can also be parallel loaded using the jam terminals P1 through P5. These terminals will program the 4018. A LOW on a jam input forces the related Q output HIGH, and vice versa. For example, if a LOW is applied to $P2$, it will force Q2 HIGH.

☐ 4022. This device is an octal, or divide-by-eight, counter that provides 1-of-8 decoded outputs. The 4022 is very nearly the same as the 4017, which is a decade version (see the discussion for the 4017).

☐ 4026. The 4026 is a decade counter that produces uniquely decoded outputs for seven-segment displays. The 4026 is a positive-edge triggered device that is fully synchronous. This chip is similar to the 4017, in that it provides an F/10 output in addition to the seven-segment decoded outputs. The decoded outputs are HIGH for *active*.

There are two *enable* inputs. One is a *clock enable* input, which will cause the count to cease when brought HIGH. The counter outputs, however, remain in their present state when the clock is inhibited. The other enable input is a *display enable* terminal. A HIGH on this input will turn the display on, and a LOW will turn the display off. A possible use of this terminal is to turn off the high-current display when it is not needed.

☐ 4029. The 4029 is an up-down counter that will divide by either 10 or 16, depending upon whether pin No. 9 is HIGH or LOW. A HIGH on pin No. 9 causes the 4029 to be a base-16 binary counter. A LOW causes it to be a base-10 decade counter.

The count direction (up/down) is determined by the level applied to PIN No. 10. If pin No. 10 is HIGH, then the 4029 operates as an up-counter, but if pin No. 10 is LOW, then it operates as a down-counter.

Chapter 10
Display Devices and Decoders

People require certain types of displays to make the data presented by digital electronic circuits intelligible. Most people, it seems, are addicted to decimal readouts, and many find binary readouts almost unintelligible.

SIMPLE BINARY READOUT

A single incandescent lamp or a light-emitting diode (LED) can form a one-bit binary readout. In most cases, the LED/lamp will be driven through an inverter (see Fig. 10-1A), or a transistor (Fig. 10-2B).

In the case of Fig. 10-1A, a TTL open-collector inverter, such as the 7405 or 7406 devices, is used to drive an LED indicator. A series resistor between the LED anode and the V+ supply limits the current to a value that can be handled safely by the LED.

When the input signal is LOW, the output of the inverter is HIGH. As a result of this condition, the LED will be turned off (LOW = off). If the input is HIGH, the output of the inverter will be LOW, or grounded). A current will flow through the LED, causing it to be lighted. In other words, HIGH = on.

In some cases, a single NPN transistor (2N3904, 2N3393, etc.) is used as the switch for the LED. The V+ supply is usually +5 VDC, but it can be any reasonable voltage. With appropriate transistor types and voltage sources, the circuit of Fig. 10-1B can also accommodate ordinary incandescent lamps, high-current incandescent lamps, and neon glow lamps (NE-2, NE-51, etc.).

The operation of the transistor driver is very straightforward:

☐ If the input is LOW (grounded), Q1 is cut off, so no current can flow through the LED (LOW = off).

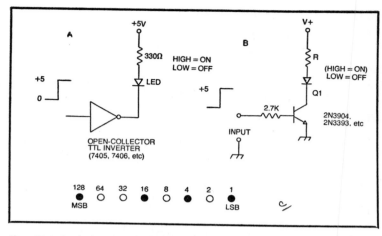

Fig. 10-1. Logic level detector using 7405/7406 inverter at A, level detector using transistor at B, and binary display.

□ If the input is HIGH, then Q1 is turned on hard; it is saturated. This places the cathode end of the LED at near-ground potential (HIGH = on).

N-BIT BINARY READOUT

The simple binary output indicators of Fig. 10-1 can be extended to multiple-bit binary circuits. We need one driver per Figs. 10-1A and 10-1B for each bit of the binary number that we wish to create. In Fig. 10-1C, an eight-bit binary display has each LED representing a single, weighted, binary digit. The normal 1, 2, 4, 8, 16, 32, 64, 128 sequence is used. In this case, the 1, 4, 16, and 128 bits are lighted (HIGH), so the binary number being represented is 10010101_2, or (128 + 16 + 4 + 1) = 149_{10}. This type of display was used on primative digital electronic instruments and is still used to display on the front panel data in certain registers of minicomputers and microcomputers.

SIMPLE DECADE DISPLAYS

Most noncomputer electronic instruments use a *decimal* display instead of a binary display. This makes the data more readable.

Figure 10-2A shows a decade adaptation of the simple binary display. Recall that many decade counters use four J-K flip-flops and have outputs weighted 1-2-4-8. If a driver/LED combination is

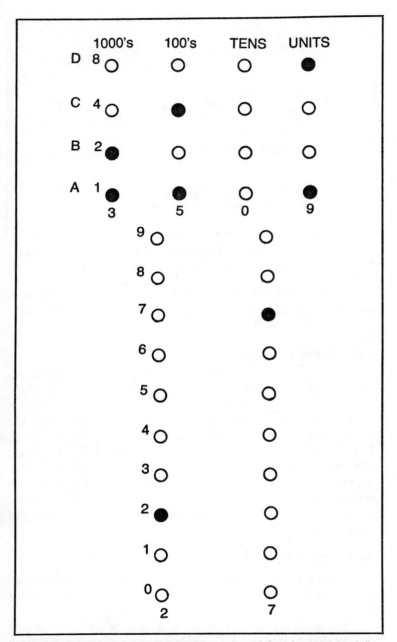

Fig. 10-2. Decade binary-weighted display at A, and decade unweighted display at B.

placed at each output bit, it will denote whether the bit is HIGH or LOW. By arranging all four bits of each decade in vertical columns, as in Fig. 10-2A, the count can be decoded. In the example, the MSD (1000s) reads 0011_2 (3_{10}), the 100s decade reads 0101_2 (5_{10}), the 10s decade is 0000_2 (0_{10}), while the units decade is 1001_2 (9_{10}). The count, then, is 3509_{10}.

An alternative column-type readout is shown in Fig. 10-2B. In this case 10 lamps represent the 10 decimal digits 0 through 9. In the two-digit case shown, the tens digit is 2 and the units is 7. The count, therefore, is 27.

The BCD display of Fig. 10-2A could be driven directly by the Q outputs of the counter flip-flops. Only a driver, such as shown in Fig. 10-1, is needed to interface the counter stage and the LEDs or lamps. The decade arrangement of Fig. 10-2B requires a *decoder* circuit, however. In this case, when interfacing a standard four-bit BCD counter, the decoder will be described as a BCD-to-one-of-10 decoder. This means that there are 10 outputs, and that one will become active for each of the 10 BCD digits. An equivalent circuit is shown in Fig. 10-3. It is in the form of a single-pole—ten-position (SP10T) rotary switch. The decoder in a real digital circuit, of course, would be an integrated circuit device. More on

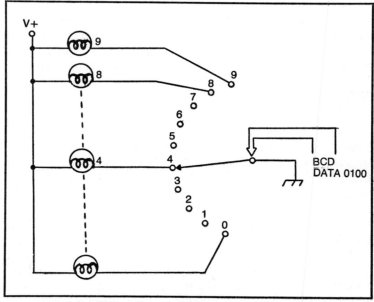

Fig. 10-3. BCD-to-1-of-10 decoder analogy.

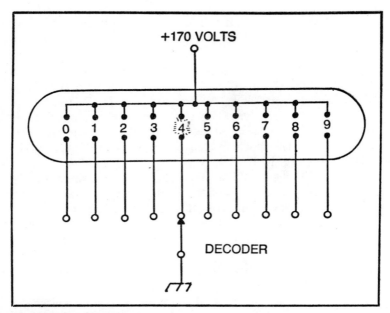

Fig. 10-4. Nixie® tube display.

this follows. In effect, however, they do the same job as the rotary switch of Fig. 10-3.

If the BCD data at the four-bit inputs of the decoder is 0100_2 (4_{10}), then decoder output No. 4 will be grounded. This will cause LED No. 4 to be lighted. Other types of displays require different decoders. Only the Nixie® tube uses the 1-of-10 decoder.

NIXIE TUBES

The *Nixie* tube, made by the Burrough's Corporation, was among the first "people-engineered" digital displays. People are trained from earliest childhood to use the Arabic digits of our decimal system. We all find it difficult to decode columns of lights rapidly in our brains, but any display that produces an Arabic digit is almost instantly recognized.

The Burrough's Nixie tube produces an Arabic numeral corresponding to the BCD word applied to the input of a Nixie decoder circuit or IC decoder. The decoder is a high-voltage version of the 1-of-10 circuit.

A Nixie tube is a neon glow lamp in which the cathodes are wire filaments formed into the various shapes of the 10 digits. A high voltage of 170 VDC is applied to a common anode terminal, so when any of the individual filaments are grounded, the neon gas particles close to their surfaces will glow orange, thus outlining the digit. The orange light emitted will take the shape of the digit. In the example shown in Fig. 10-4, the "4" cathode is grounded, so the digit lighted will be 4.

The Nixie tube suffers from two major defects: parallax and high voltage. The parallax problem exists because all 10 digits inside the tube are in slightly different planes. Although Fig. 10-4 shows the digits spread out, they are in reality placed one behind the other, so that they can be viewed from the front. This causes viewers who are positioned at an angle to see a distorted view. Typically, some digits will be seen fully, while others are distorted almost to the point of invisibility.

The Nixie tube has been all but eclipsed by more modern devices in newly designed equipment. Many instruments on the market still use Nixie tubes, however.

SEVEN-SEGMENT READOUTS

The seven-segment readout is probably the most common type of display used in digital electronics. It is used in calculators, frequency/period counters, and most other forms of instruments.

The basic seven-segment readout is shown in Fig. 10-5A and consists of seven lighted bars, or segments, labeled a through g.

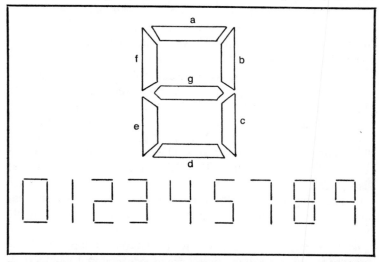

Fig. 10-5. Seven-segment readout at A, and seven-segment versions of the decade digits.

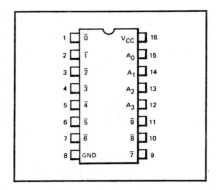

Fig. 10-6. 7442 decoder.

The decoder will cause appropriate segments to be lighted, forming the numeral. The light patterns for the 10 decimal digits are shown in Fig. 10-5B. Certain alphabet characters can also be represented, but only a few are easily recognized with the rest being quite poor representations of the actual letters. The 5×7 dot matrix of the next section will represent alphabetic characters more easily.

The seven-segment readout is available using several different illumination technologies. The RCA Numitron, for example, uses seven incandescent filaments to form the segments. These tubes are available in two sizes. The larger is housed in a nine-pin version of the small (usually 7-pin) vacuum tube envelope. The larger is in a standard 9-pin vacuum tube envelope.

The most common seven-segment display device uses LEDs to form the lighted segments. These are the red displays seen on so many electronics watches, calculators, etc.

The LED type does not usually offer the necessary brightness for viewing across the room or in a too-well lighted area (try viewing an LED display in sunlight). The gas discharge seven-segment display is brighter and can be larger, but only at the expense of requiring 150 to 200 VDC. This type of display produces a bright orange display color. On the other hand, the fluorescent display requires 20 to 40 VDC and produces a blue-green display color.

The display types mentioned above require substantial amounts of electrical current, up to 50 milliamperes per segment. If an "8" is illuminated, using all seven segments, it would draw as much as 8×50 mA, or 400 mA, from a DC power supply—all for one digit. An eight-digit digital frequency counter would then require 8×400 mA, or 3.2A, to display the frequency 88888888.

The liquid crystal display, however, uses current conservatively. These displays are the dark gray-on-light gray types. Some use a yellow filter to provide a yellowish background. This improves the contrast for the viewer.

Liquid-crystal displays are *electrostatic* devices that only draw current when *changing state*. The static current drain is so small that CMOS calculators using liquid-crystal displays can operate as long as 1500 hours (one year of normal use) on a single pair of hearing aid batteries.

DECODERS

The binary or BCD code produced by a digital counter is not capable of directly driving the decimal display device. Some form of *decoder* circuit is needed to convert the four-bit binary, or BCD, to the one-of-ten or seven-segment codes used by the displays.

All of the decoders can be made to use assorted gates, properly interconnected, but it is far easier to use some of the various TTL and CMOS IC decoders that are available.

The 7442

The TTL device shown in Fig. 10-6 is a BCD-to-1-of-10 decoder. The 7442 outputs remain HIGH, except for the one selected by the BCD code applied to the inputs. The *selected* output goes LOW. The 7442 outputs can sink up to 16 mA of current at

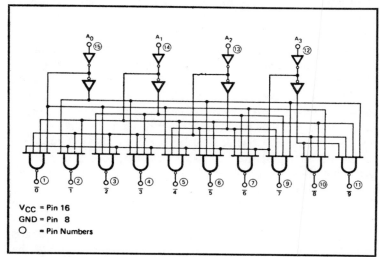

Fig. 10-7. 7442 decoder schematic.

141

TTL voltage levels. A companion device, the 7445, will sink up to 80 mA and operates at voltages up to 30 VDC (output terminals only—the rest of the circuits inside the 7445 require the +5 VDC normally used in TTL).

The pinouts for the 7442/7445 are shown in Fig. 10-6, while the equivalent internal circuit is shown in Fig. 10-7. The internal operation is not very important to your understanding; it is included here merely as an example of how decoding can be accomplished. No other internal circuits will be shown, unless there is a compelling reason to do so.

The 7447, 7448, and 7449

These TTL devices operate as BCD-to-seven-segment decoders. The pinouts for the 7447 are shown in Fig. 10-8A, while those for the 7448 and 7449 are shown in Figs. 10-8B and 10-8C, respectively.

Like the previous example, an active output goes LOW when the appropriate BCD input word is applied. Unlike the 7442/45, however, a seven-segment decoder may have more than one output LOW at any given time. For example, if the BCD code 0110_2 were applied to the inputs, decimal "6" ($0100_2 = 6_{10}$) would be displayed. This means that the outputs controlling the c, d, e, f, and g segments of the seven-segment display would have to be LOW.

The 7447 can sink up to 40 mA of current at each output, and the outputs can handle up to +30V. The package supply, however, must be +5 VDC.

The 7447 may be used directly on incandescent and fluorescent seven-segment displays, provided that the output current and voltage requirements are met. If LED seven-segment displays are used, however, a current limiting resistor must be connected in series with each output. If a V+ supply of +5 VDC is used, then 330-ohm resistors are used. As higher voltages are used, though, proportionally higher resistances must be used.

The test, or LT, terminal on the 7447 is used to provide a test of the decoder and the seven-segment display. If this pin is grounded, the display will read 8. If you examine Fig. 10-5B again, you will see than an 8 requires that all seven segments be lighted. This feature allows a quick visual test of the system in order to rapidly spot dead segments.

The RBI and RBO terminals are the *ripple blanking input* and *ripple blanking output*, respectively. These are used to turn off the display in order to eliminate leading zeroes in a multidigit display.

142

If the blanking input is LOW, the display is turned off whenever the BCD code 0000_2 is applied to the inputs. The blanking output of the previous stage provides the LOW condition that turns off the display.

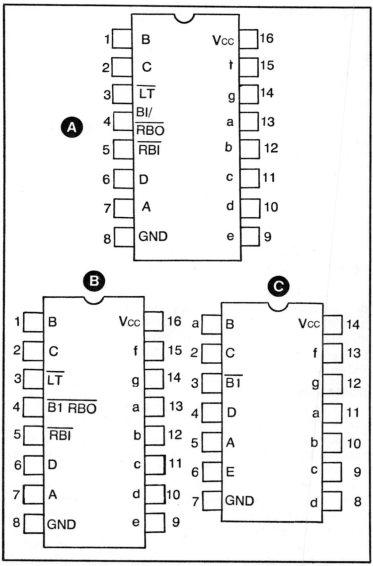

Fig. 10-8. 7447 decoder at A, 7448 decoder at B, and 7449 decoder at C.

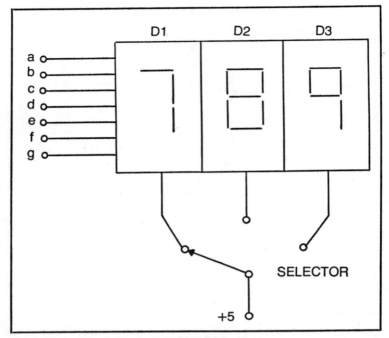

Fig. 10-9. Electrical analogy for a multiplexed display.

The 7447 is probably the most commonly used seven-segment decoder in the TTL line. The 7446 device is similar to the 7447, but uses open-collector outputs that require a pull-up resistor or segment load. The 7448 uses internal 2000-ohm pull-up resistors and has a pinout similar to the 7447. It will, however, sink less current and must operate at +5 VDC.

DISPLAY MULTIPLEXING

The display device in any digital instrument may require more current than the rest of the electronic devices in the circuit put together. One incandescent seven-segment display, for example, requires 40 mA per segment. Consider an eight-digit frequency counter in which the count is "88888888" (all seven segments on all eight readouts turned on). The current requirement is $(8 \times 7 \times 40)$ mA = 2240 mA = 2.24 amperes! Clearly, then, digital readouts eat up a lot of current. If an instrument is to be battery-powered, then the display severely limits operating time.

One answer is to use *display multiplexing* to reduce the total *on-time* of each display device. This saves a large fraction of the energy that a steady-state display would consume.

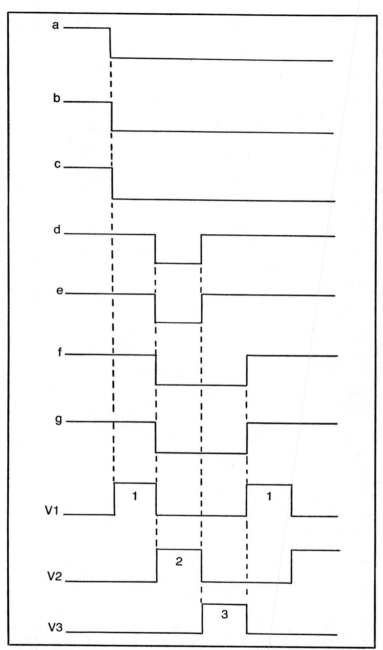

Fig. 10-10. Timing diagram of a multiplexed display.

145

Multiplexing is a technique whereby each digit is turned on in sequence, with all others being off. If the switching is done fast enough, such as at a 200-kHz rate, each display will be on only part of the time, yet the display *seems* to remain on all of the time. The persistence of the viewer's eye—the latent image retained for a split second after the light actually extinquishes—makes the display appear steady.

A simplified three-digit multiplexed display is shown in Fig. 10-9, while the timing diagram is shown in Fig. 10-10. Three seven-segment displays are shown in Fig. 10-9. The seven control lines to the decoders, on all seven lines, are connected together to form a seven-line *bus*. This means that all three lines from the *a* segments are connected together to form the *a* bus line, all of the *b*'s, and all of the *c*'s, and so on.

When a seven-segment code is applied to the bus, it is applied simultaneously to all three digits. Only one digit lights up at a time, however, because the +5 VDC supply is connected to only one digit at a time.

An electronic selector circuit determines which digit is lighted. This circuit can be viewed as an electronic rotary switch that applies +5 VDC to only one digit at a time. In Fig. 10-10, lines V1 through V3 are the V+ lines (+5V) to digital display devices D1 through D3, respectively.

Examine the timing diagram shown in Fig. 10-10. Let's say you want to display the decimal number "784." When +5V is applied to V1, making V1 is HIGH, the counter outputs the code for 7. In other words, *a*, *b*, and *c* are LOW and all others are HIGH. Although all three digits see this same code, only V1 can respond at this time, because it is the only display device with +5 VDC applied.

When V1 goes LOW and V2 goes HIGH, digit 2 will become energized. At this time the counter outputs the seven-segment code for 8. This means that all segment outputs are LOW. The second digit now displays an 8 and the others are off. When V3 goes HIGH (V1 and V2 are now LOW), the counter output code is for 4 (*b*, *c*, *f*, and *g* are LOW), so D3 lights up and displays a 4.

The example of Fig. 10-9 is for a three-digit circuit. In a calculator or counter, however, 8 to 12 digits may be displayed simultaneously. These longer displays must be cycled through at a faster rate than a three-digit display, or the readout will appear to be dimmed; in a 12-digit display, each seven-segment readout device is turned on only one-twelfth of the time.

Chapter 11
Registers

A flip-flop is able to store a single *bit* of digital data. When two or more flip-flops are organized into some configuration to store multiple bits of data, they constitute a *register*. Most registers are merely specialized arrays of ordinary flip-flops.

There are several different circuit configurations that one would call registers. We classify these according to the manner in which data is input to and output from the register. We have, for example, *serial-in-serial-out* (SISO), *serial-in-parallel-out* (SIPO), *parallel-in-parallel-out* (PIPO), and *parallel-in-serial-out* (PISO) registers.

SISO AND SIPO

Figures 11-1 and 11-2 represent both SISO and SIPO shift registers. The only really significant difference is the parallel output lines used on the SIPO device; these lines would be absent on the SISO register. Note that some registers are both SISO and SIPO, depending upon how you use them. The SIPO shift register consists of a cascade chain of type-D flip-flops, with the clock lines tied together; i.e., they share a common clock line.

Recall one of the rules for Type-D flip-flops: Data can be transferred from the D-input to the Q-output *only* when the *clock* input is HIGH. This rule can be applied to the situation shown in Fig. 11-2, where the transmission of a single data bit from left to right through the SISO shift register is shown.

At the occurence of the first clock pulse, the input is HIGH. This point is the D-input of FF1, so a HIGH is transferred to the Q1 output. This HIGH, which is also applied to the D input of the second flip-flop (FF2), remains after the clock pulse vanishes.

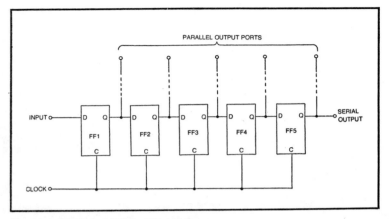

Fig. 11-1. SIPO and SISO shift register.

When the second clock pulse arrives, FF2 sees a HIGH on its D-input, and FF1 sees a LOW on its D-input. This situation causes a LOW at Q1 and a HIGH at Q2.

The third clock pulse sees a LOW on the D-inputs of both FF1 and FF2, and a HIGH at the D-input of FF3. The third clock pulse, then, causes Q1 and Q2 to be LOW and Q3 to be HIGH.

Note that the SISO input remains LOW after the initial HIGH during clock pulse No. 1. This means that the single HIGH will be propagated through the entire SISO shift register, one stage at a time. The HIGH bit will shift one flip-flop to the right each time a clock pulse arrives. If the data at the input had changed, then the bit pattern at the input will be propagated through the shift register.

The shift register shown in Fig. 11-1 is a five-bit, or five-stage, shift register, although any bit-length could be selected. On the sixth clock pulse, therefore, the HIGH is propagated out of the register, so all flip-flops are LOW; i.e., $Q1=Q2=Q3=Q4=Q5=\emptyset$.

The SISO shift register can be made into a SIPO device by adding parallel output lines, one each for Q1, Q2, Q3, Q4, and Q5. One use for the SIPO shift register is serial-to-parallel code conversion. For economic reasons, digital data is usually transmitted from device to device as a serial stream of bits, especially if long distances are involved. In other words, the bits of the digital word are sent over a communications channel one at a time. But most computers and other digital instruments use data in the more efficient parallel form. This format is both faster and more expensive; hence, the use of serial in transmission. As an example, consider an eight-bit binary data system. We would need an eight-

148

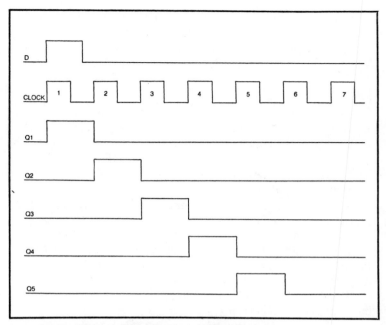

Fig. 11-2. Timing diagram of SIPO and SISO shift register.

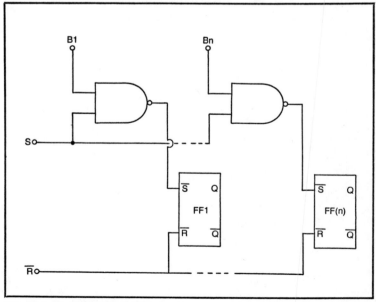

Fig. 11-3. Parallel entry shift register.

stage SIPO to convert the eight-bit serial binary code to eight-bit parallel form. The code is entered into the SIPO register one bit at a time, so that after eight clock pulses, the first bit will appear at Q8, and the last bit at Q1.

PARALLEL

Parallel entry shift registers are faster to load than the serial input types because a given bit in the shift register can be changed directly, without applying it through all of the other stages. The new bit need not ripple through all of the preceding stages to reach its destination. There are two basic types of parallel entry: *parallel-direct* and *jam*.

In parallel-direct entry, or simply parallel entry, shown in the partial schematic of Fig. 11-3, the register must first be cleared by setting all bits to LOW and first bringing the *reset* line momentarily LOW. The data that is applied to inputs B1 through B_n can then be loaded into the register by momentarily bringing the *set* line HIGH.

The jam entry circuit of Fig. 11-4 is able to load data from the B1 through B_n inputs in a single operation. This job is accomplished by loading the complements of the B1 through B_n data into the other inputs. In actual IC shift registers using this technique, inverter stages are connected internally to the inputs to automatically

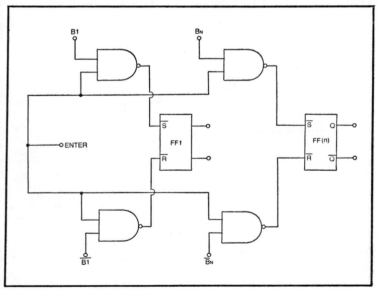

Fig. 11-4. Jam entry shift register.

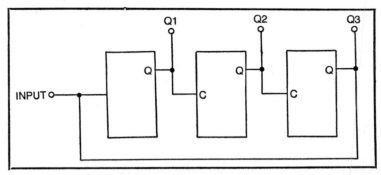

Fig. 11-5. Recirculating shift register.

complement the input lines. The outside user never sees the complementing process.

A recirculating shift register circuit is shown in Fig. 11-5. These registers are able to readout the data in serial format and then re-enter it so that it is retained. This is done by connecting the output of the SISO register back to the input. As the data is read out of an ordinary SISO register, it is destroyed and lost forever; however, we sometimes wish to retain the data and be able to read it out, too. The solution here is to read it right back in as it is read out. Although only three bits are shown here for simplicity, the bit length could be almost anything.

One interesting application for the recirculating shift registers is in non-fade medical oscilloscopes. Most of the human physiological signals that physicians want displayed on a scope are very slow; i.e., 1 Hz represents a heart rate of 1 BPM. The trace tends to fade before the entire waveform is displayed. Modern oscilloscopes overcome this problem by digitizing the ECG waveform and then storing it in a recirculating shift register. By scanning the memory, which means reading out the serially stored data to a digital-to-analog converter, the entire waveform can be displayed on the CRT screen at one time. This is done 64 to 256 times per second in commercially available models. New waveform data is usually added to the shift register in four-bit groups.

IC EXAMPLES

Few designers would bother to use a series of cascade flip-flops in making a shift register. There are simply too many different types of IC shift register available on the market. Both TTL and CMOS examples abound.

151

Chapter 12
Timers and Multivibrators

A timer is a circuit that is able to produce an output level for a precise length of time. This output level may be normally HIGH and will drop LOW for the specified period of time; or, alternatively, it could be normally LOW, and then snap HIGH for the specified period when active. Devices of both sorts are known.

Many timers are also part of a family of circuits called multivibrators. In the case of the timer described above, we are talking about a *monostable multivibrator* or one-shot stage. There are also *bistable* and *astable* multivibrators.

The names of these multivibrators are derived from the number of stable output states that each will allow. The monostable multivibrator, for example, has but one stable stage. It normally rests in its stable, or dormant, state. But when a trigger pulse is received, the output snaps to the active state. It will remain in the active state only for a given period of time, after which the output snaps back to the stable state.

The bistable multivibrator is our old friend of RS flip-flops. Recall from Chapter 7 that one trigger input on this stage will SET the output and make Q HIGH, while a trigger pulse on the other input will RESET the output or make Q LOW. The RS flip-flop, however, doesn't much care which state it is in. It would remain happily in either SET or RESET condition; two stable states exist which is where the name *bistable* multivibrator comes from.

The astable multivibrator has *no* stable states. It will merrily flip and flop back and forth between SET and RESET conditions. The astable multivibrator, therefore, functions as a *clock generator*, producing a square wave output wave train.

Both astable and monostable multivibrators function as timer circuits. Bistables, however, are not easily used in this capacity.

TIMERS VERSUS CLOCKS

It is usually the practice to distinquish *timers* from *clocks*. A timer circuit is the one-shot described earlier in this chapter. On the other hand, a clock is an astable multivibrator. Clocks are used to synchronize digital circuits.

There are several ways to generate a timer pulse. One of the most popular is the 555 integrated circuit timer. There is also a 556 device, which is a dual 555.

The 555 is inside an 8-pin miniDIP integrated circuit package. It has the advantage of being essentially free of period drift caused by variations in power-supply voltages; something that cannot be said of the unijunction transistor (UJT) relaxation oscillators previously used as timers.

The block diagram of the 555 is shown in Fig. 12-1, while the pinouts of the 8-pin miniDIP package is shown in Fig. 12-2. The 555 is one of the most popular IC devices used for timer circuits. Perhaps one of the reasons for this is that it is neither TTL nor CMOS, yet it can be interfaced with either of those logic families. By the way, the 555 is of bipolar transistor construction.

One principal difference between the 555 and TTL devices is that the 555 can be operated over a wide range of potentials; i.e., 5V to 15V, although potentials between 9V and 12V seems optimum. When the output terminal is LOW, the output terminal (pin No. 3) can *sink* up to 200 mA. When the output is HIGH, on the other hand, it will *source* up to 200 mA.

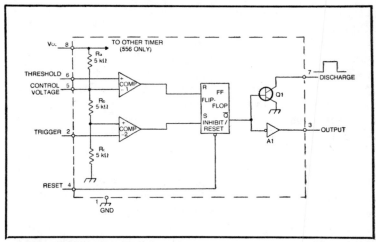

Fig. 12-1. Block diagram of 555 timer IC.

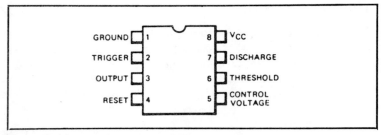

Fig. 12-2. Pinouts and package of 555.

The 555 is capable of operating in either monostable or astable modes. The major difference between these modes of operation is the use of the output to retrigger the device.

The 555 is one of those nice devices that will allow the clever designer to use a lot of imagination. Before you can fully understand the wide range of uses this chip can offer, however, it is necessary to know the inner workings intimately. For this reason, we are going to explain the modes of operation using modified versions of Fig. 12-1 in the illustrations.

Figure 12-3 shows a monostable multivibrator using the 555 IC timer. Figure 12-3 is the block diagram, while Fig. 12-4 is the circuit as it appears in a schematic diagram.

The heart of the 555 timer is an RS flip-flop (FF) that is controlled by a pair of voltage comparators. An RS flip-flop, you will recall, is a bistable circuit; that is, it has two stable states. In its initial state the Q output is LOW, and the not-Q is HIGH. If a pulse is applied to the FF-SET (S) input the situation reverses itself, and the not-Q output becomes LOW, while the Q is HIGH.

A *comparator* is a device that is capable of comparing two voltage levels, and issuing an output that indicates whether they are equal or unequal. A comparator can be made by taking an operational amplifier or other high gain linear amplifier without any negative feedback. In the 555, two comparators are designed so that their outputs go HIGH when the two inputs are at the same potential.

Under initial conditions, at time T_o, the not-Q terminal of the RS flip-flop is HIGH, and this biases transistor Q1 hard on, placing pin No. 7 of the 555 at ground potential. This keeps capacitor C1 discharged. Also, amplifier A1 is an inverter, so the output terminal (pin No. 3) is initially in a LOW condition.

Resistors R_a, R_b, and R_c are inside the 555 IC and are of equal value, nominally 5000 ohms. These form a voltage divider that is used to control the voltage comparators. The inverting (-) input of comparator No. 1 is biased to a potential of:

$$E_1 = \frac{(V_{cc})\,(R_b + R_c)}{R_a + R_b + R_c}$$

$$E_1 = \frac{2}{3}\ V_{cc}$$

This means that the output of comparator No. 1 will go HIGH when the control voltage to IC pin No. 5 is equal to $2/3\text{-}V_{cc}$. Similarly, the same voltage divider is used to bias comparator No. 2. The voltage applied to the noninverting input of the second comparator is given by:

$$E_2 = R_a + R_b + R_c \times \frac{R_c}{V_{cc}}$$

$$= \frac{1}{3} \times V_{cc}$$

When the voltage applplied to the trigger input (IC pin No. 2) drops to $\frac{1}{3}\text{-}V_{cc}$, the output of the second comparator goes HIGH.

The control voltage is, in this case, the voltage across capacitor C2. This capacitor charges through resistor R_a and will reach $2/3\text{-}V_{cc}$ in less than 1 ms after power is applied. If a short,

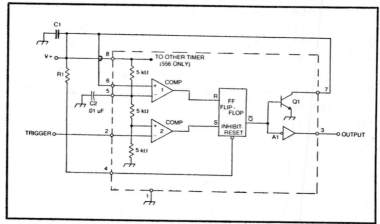

Fig. 12-3. 555 in a monostable circuit.

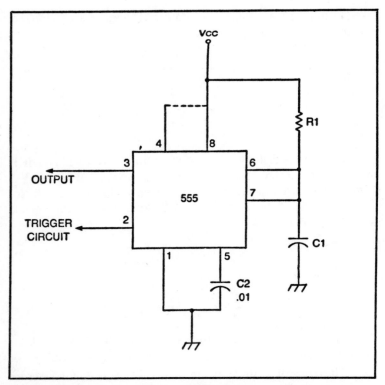

Fig. 12-4. Same circuit as seen in circuit diagrams.

negative-going pulse is applied to the trigger input at time T_1 in Fig. 12-4, then the output of comparator No. 2 will drop LOW as soon as the trigger pulse voltage drops to a level equal to $1/3\text{-}V_{cc}$. This puts the flip-flop in the SET condition and causes the not-Q output to drop to the LOW state.

A drop to the LOW state by the not-Q output at time T_1 causes two things to occur simultaneously: One is to force the output of buffer amplifier A1 HIGH, and the other is to turn off transistor Q1. This allows capacitor C1 to begin charging through resistor R1. The voltage across C1 is applied to the noninverting input of comparator No. 1 through the threshold terminal (pin No. 6). When this voltage rises to $2/3\text{-}V_{cc}$, comparator No. 1 will toggle to its HIGH state and will RESET the RS flip-flop. This occurs at time T_2 and forces the not-Q output again to its HIGH state.

The output of amplifier A1 again goes LOW, and transistor Q1 is turned on again. When Q1 is on, we find that capacitor C1 is discharged. At this point the cycle is complete, and the 555 timer is again in its dormant state. The output terminal will remain LOW until another trigger pulse is received. The approximate length of time that the output terminal remains in the HIGH condition is given by:

$$T = T_2 - T_1$$
$$T = 1.1 \text{ R1C1}$$

where T is the time duration in seconds, R1 is the resistance of *R1* in ohms, and C1 is the capacitance of *C1* in farads.

This function is graphed in Fig. 12-5 for times between 0.01 and 10.0 seconds with values of R1 and C1 that are easily obtainable. The time relationship between the trigger pulse, the output pulse, and the voltage across capacitor C1 is shown in Fig. 12-6.

If the reset terminal is not used, it should be tied to V_{cc} to prevent noise pulses from jamming the flip-flop. If, however, negative-going pulses are applied simultaneously to the trigger input (pin No. 2) and the reset terminal (pin No. 4), the output pulse will terminate. When this occurs, the output terminal drops back to the LOW state, even though the time duration (T) has not yet expired.

An astable multivibrator is similar in many respects to the monostable variety, except that it is self-retriggering. The astable multivibrator has no stable states, so its output will swing back and forth between the HIGH and LOW states. This action produces a wave train of square wave pulses. An example of a 555 astable

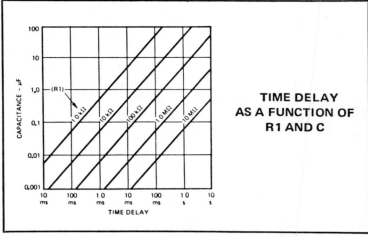

Fig. 12-5. Operating parameters of the 555 in monostable operation.

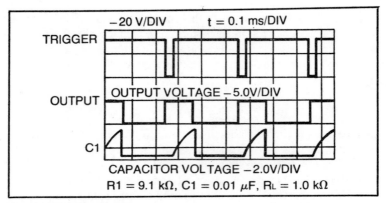

Fig. 12-6. Waveforms of the 555.

$$T = T_1 + T_2$$
$$T = 0.693\ (R1 + 2R2)\ C1$$

multivibrator circuit is shown in Figs. 12-7 through 12-9. Again, we have a version of Fig. 12-1 for block diagram analysis in Fig. 12-7, and the circuit as it will appear in schematic diagrams is shown in Fig. 12-8. As in the previous case, the inverting input of comparator No. 1 is biased to $2/3\text{-}V_{cc}$, and the noninverting input of comparator No. 2 is biased to a level of $1/3\text{-}V_{cc}$ through the action of resistor voltage divider network R_a, R_b, and R_c. The remaining two comparator inputs are strapped together and are held at a voltage determined by the time constant $C1(R1 + R2)$. Under initial conditions, the not-Q output of the RS flip-flop is HIGH. This turns on transistor Q1, keeping the junction of resistors R1 and R2 at ground potential. Capacitor C1 has been charged, but when Q1 turns on, C1 will discharge through resistor R2. When the voltage across capacitor C1 drops to a level of $2/3\text{-}V_{cc}$, the output of comparator No. 1 goes HIGH and that resets the flip-flop. This action again turns on Q1 and allows C1 to discharge to $1/3\text{-}V_{cc}$. Capacitor C1, then, alternatively charges to $2/3\text{-}V_{cc}$ and then discharges to $1/3\text{-}V_{cc}$. Figure 12-9 shows the relationship between HIGH and LOW times. The high time, T_1, is given by:

$$T_1 = 0.693\ (R1 + R2)\ C1$$

and the low time by:

$$T_2 = 0.693\ (R2)\ C1$$

The total period of the waveform, T, is the sum of T_1 and T_2, and is given by:

158

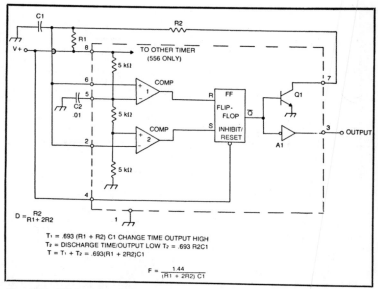

Fig. 12-7. 555 in astable operation.

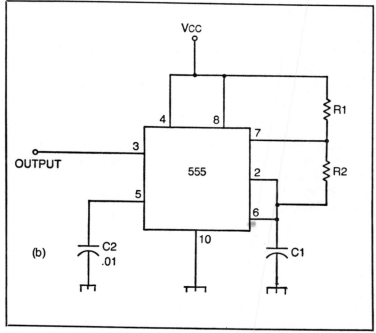

Fig. 12-8. Same circuit as seen in circuit diagrams.

159

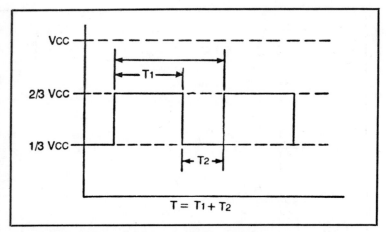

Fig. 12-9. Timing diagram of 555 in astable operation.

In any electrical circuit, or any other physical system for that matter, the frequency of an oscillation is the reciprocal of the period. In this case, then, the frequency of oscillation is:

$$F_{Hz} = 1/T_{sec}$$

$$F_{Hz} = \frac{1}{0.693\ (R1 + 2R2)\ C1}$$

$$F_{Hz} = \frac{1.44}{(R1 + 2R2)\ C1}$$

The last equation is shown graphically in Fig. 12-10 for frequencies between 0.1 Hz and 100,000 Hz, using easily obtainable component values. The relationship between the C1 voltage and the output state is shown in Fig. 12-11. The *duty cycle*, also called *duty factor*, is the percentage of the total period that the output is HIGH. This is given by:

$$DF = \frac{T_1}{T_1 + T_2}$$

$$DF = \frac{R2}{R1 + R2}$$

The trigger input of the 555 timer is held in the HIGH condition in normal dormant operation. To trigger the chip and initiate

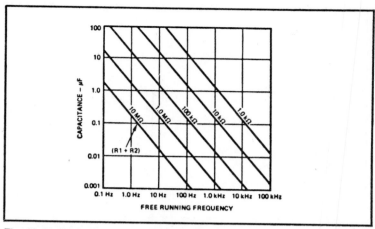

Fig. 12-10. Parameters of the 555 in astable operation.

the output pulse pin No. 2 must be brought down to a level of $1/3\text{-}V_{cc}$. Figure 12-12 shows how this can be done manually. Resistor R4 is a pull-up resistor used to keep pin No. 2 HIGH. Capacitor C3 is charged through resistors R3 and R4. When S1 is pressed, the junction of R3 and C3 is shorted to ground, rapidly discharging C3. The sudden decay of the charge on C3 generates a negative-going pulse at pin No. 2 that triggers compatator No. 2, initiating the output pulse sequence.

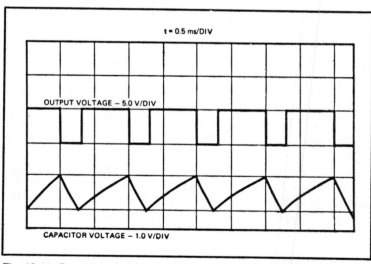

Fig. 12-11. Operating waveforms of the 555 in astable operation.

161

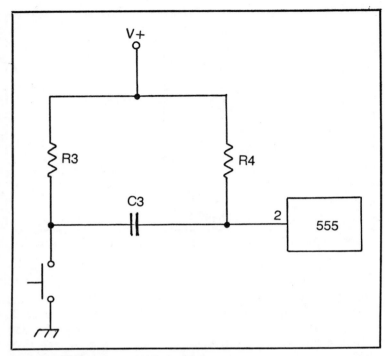

Fig. 12-12. Manual triggering of the 555.

The 555 timer is also available in a dual version called the 556. This IC is a 14-pin DIP package containing two timers, independent of each other except for the power-supply voltage. It is, then, a dual 555. Both 555 and 556 timers are so useful that it is an even bet you will someday use one or both if you do any circuit design work at all.

Another useful IC timer—in fact, a *family* of timers—is the Exar type XR-2240, XR-2250, and XR-2260. The internal circuitry of these is shown in Fig. 12-13. These devices are second sourced in the form of the Intersil 8240, 8250, and 8260, respectively. The 2240 is a binary counter with 8-bit outputs. The 2250 is a similar device, except that the outputs are in BCD code. The 2260 is also BCD, except that the most significant bit is 40, instead of 80. This allows the 2260 to be used as a timer with 60 count rather than 100. In the 2250 and 2260 devices, the regulator output terminal (pin No. 15) is used as an overflow that allows cascading of several stages. Table 12-1 shows the pinouts and their respective weighting.

162

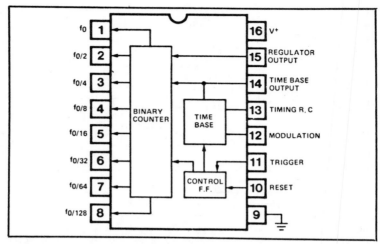

Fig. 12-13. Pinout diagram of the XR-2240 timer.

These timers will operate over a supply voltage range of +4.5 to +18 VDC. The time base section is a clock circuit that is very similar to the 555 device. This similarity is shown in Fig. 12-14, which shows the internal circuitry of the XR-2240. One main difference between the 555 timer and the time base portion of the XR-2240 lies in the relative reference levels created by the respective internal voltage dividers (R1, R2, and R3 in Fig. 12-14). In the 555 timer all three resistors were of equal value, so comparator input terminals were at $0.33V_{cc}$ and $0.66V_{cc}$. In the Exar chip, on the other hand, the reference levels are $0.27V_{cc}$ and $0.73V_{cc}$, respectively. One result of this is simplification of the duty factor equation that gives the period of the waveform.

The binary counter section consists of a chain of JK flip-flops connected in the standard manner, where each stage functions as a

Table 12-1. Pinouts of the Exar Timers.

Pin No.	2240/8240	2250/8250	2260/8260
1	1	1	1
2	2	2	2
3	4	4	4
4	8	8	8
5	16	10	10
6	32	20	20
7	64	40	40
8	128	80	(N.C.)

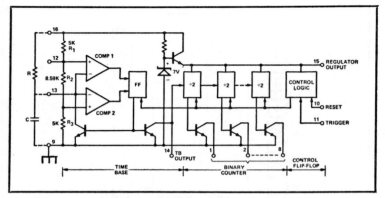

Fig. 12-14. Block diagram of the XR-2240 timer.

divide-by-2, or binary, counter. The binary counter chain is connected to the output of the time base section through an NPN open-collector transistor. The transistor collector is also connected to IC pin No. 14 (called *time base output*) so that a 20K ohm pull-up resistor can be connected between the collector and the output of internal regulator pin No. 15.

Digital outputs from this counter are, in the usual fashion, given as voltage levels at a set of IC pins. Each output bit is delivered to a specific terminal of the IC package where it is connected to a pull-up resistor similar to that used for the time base output terminal. The output terminals will generate a LOW condition when active. This may seem to be opposite to the usually accepted arrangement, but there is a method to this madness, and it does create a highly versatile, stable, long-duration counter. These are properties not usually associated with single-IC designs.

Figure 12-15 shows the basic operating circuit for the XR-2240 timer IC. This chip proves interesting because the sole difference between circuits for astable and monostable operation is the 51K ohm feedback resistor linking the reset terminal (pin No. 10) to the wired-OR outputs. The timer is set into operation by application of a positive-going trigger pulse to pin No. 11. This pulse is routed to the control logic and has several jobs to perform simultaneously: resetting the binary counter flip-flops, driving all outputs low, and enabling the time base circuit. As was true in the 555 IC, this timer works by charging capacitor C1 through resistor R1 from a positive voltage source, V+ or V_{cc}. The period of the output waveform is given by:

$$T = R \times C$$

where T is the period in seconds, R is the resistance of R1 in ohms, and C is the capacitance of C1 in farads.

The pulses generated in the time base section are counted by the binary counter section, and the output stages change states to reflect the current count. This process will continue until a positive-going pulse is applied to the reset terminal.

Figure 12-16 shows the relationship between the trigger pulse, time base pulses, and various output states. The reason for the open-collector output circuit is to allow the user to wire a permanent OR output so that the actual output duration can be programmed. Each binary output is wired in the usual power-of-two sequence: 1, 2, 4, 8, 16, 32, 64, and 128. If these are wired together, the output will remain LOW as long as *any one output* is LOW. This allows the output duration to be programmed from 1T to 255T, where T is defined as in the last equation, by connecting together those outputs which sum to the desired time period. For example, design a timer with a 57-second time delay. In the binary notation, decimal 57 is equal to:

$$32 + 16 + 8 + 1 = 57$$

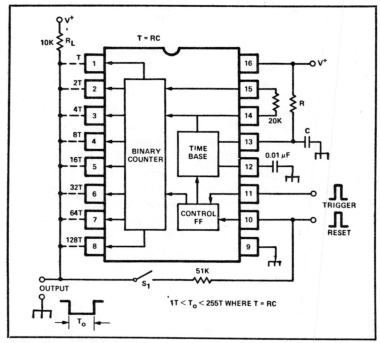

Fig. 12-15. XR-2240 circuit.

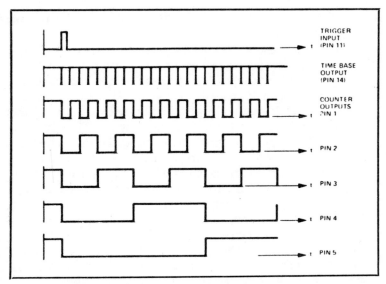

Fig. 12-16. Timing diagram of the XR-2240 timer.

Set $T = R1C1 = 1$ second wire and wire together the pins on the XR-2240 corresponding to these weights, then a 57-second time delay will be realized. The base diagram to this IC shows us that these pins are numbers 1, 4, 5, and 6. If those four pins are shorted together, the counter output will remain LOW for 57 seconds following each trigger pulse.

The time base or the wired-OR terminals could be changed to vary the output periods. Of course, if the time base frequency were doubled, the counter would reach the desired state in half the time. This feature allows programming of the XR-2240 to time durations that might prove difficult to achieve using conventional circuitry.

Each output must be wired to V_{cc} through a pull-up resistor of 10K ohms, unless, of course, the wired-OR output configuration is used. In that case, a single 10K ohm resistor is used.

Current through the output terminals must be kept at a level of 5 mA or less. This serves as a general guide to the selection of pull-up resistors.

The amplitude of the reset and trigger pulses must be at least two PN junction voltage drops ($2 \times 0.7 = 1.4V$). In most practical applications, it might be wise to use pulses greater than 4V amplitude, or standard TTL levels, in order to guard against the possibility that any particular chip may be a little difficult to trigger

166

near minimum values or that outside factors tend to conspire to actually reduce pulse amplitude at the critical moment.

Synchronization to an external time base or modulation of the pulse width is possible by manipulating pin No. 12. In normal practice, this pin, which is the noninverting input of comparator No. 1, is bypassed to ground through a 0.01 μF capacitor so that noise signals will not interfere with operation. A voltage applied to pin No. 12 will vary the pulse width of the signal generated by the time base. This voltage should be between +2V and +5V for a time base change multiplier of approximately 0.4 to 2.25, respectively.

If you wish to synchronize the internal time base to an external reference, connect a series RC network consisting of a 0.1 μF capacitor and a 5.1K ohm resistor to IC pin No. 12. This forms and input network for sync pulses, and these should have an amplitude of at least 3V at periods between 0.3T and 0.8T (see Fig. 12-17). Another way to link the count rate to an external reference is to use an external time base. This signal may be applied to the *time base output* terminal at pin No. 14.

Each Exar XR-2240 has its own internal voltage regulator circuit to hold the DC potentials applied to the binary counters at a level compatible with TTL logic. This consists of a series pass transistor, which has its base held to a constant voltage by a zener diode. If operation below 4.5 VDC is anticipated, it becomes necessary to strap the regulator output terminal at pin No. 15 to V_{cc} (V+) at pin No. 16. The regulator terminal can be used to source up to 10 mA to external circuitry or an additional XR-2240.

LONG-DURATION TIMERS

Long-duration timers can be built using any of several approaches. You could, for example, connect a unijunction transistor

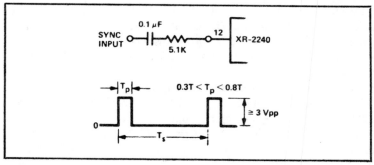

Fig. 12-17. Sync input.

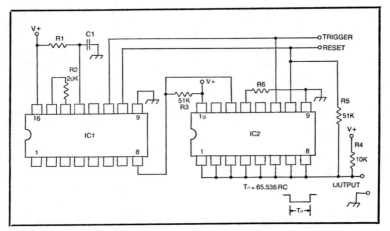

Fig. 12-18. Cascading to make a long-duration timer.

(UJT) in a relaxation oscillator configuration, or use a 555. In almost all cases, though, there seems to be an almost inevitable error created by temperature coefficient and inherent tolerance limits of the high-value resistors and capacitors required. Also, a certain amount of voltage drop is across the capacitor due to its own internal leakage resistances and the impedance of the circuit in which it is connected. The use of the XR-2240 IC timer will all but eliminate such problems because a higher clock frequency is allowed. Easier-to-tame component values will therefore satisfy the equation, $T = R \times C$. It is usually easier to specify and obtain high-quality, precision components in these lower values. This makes the period initially more accurate and results in less drift with changes in temperature.

An example of a long-duration timer is shown in Fig. 12-18; it consists of two XR-2240 IC timers cascaded to increase the time duration. In this circuit, the time base output of IC2 (pin No. 14) is an input for an external time base. Timer IC1—also an XR-2240— is used as a time base for IC2. The most significant bit of IC1 (pin 8, weighted 128) is connected to pin No. 14 of IC2. This pin will remain LOW from the time IC1 is triggered until time $T_o = 128R1C1$. It will then go HIGH and trigger IC2.

The binary counters in IC2 will increment once for every 128T. Timer IC1 is essentially operating in the astable mode because its reset pin is tied to the reset of IC2, and that point does not go HIGH until the programmed count for IC2 forces its output to go HIGH.

The total time duration for this circuit under the conditions shown (IC2 input from pin No. 8 of IC1, and with all IC2 outputs connected into the wired-OR configuration) is 256^2, or 65,536T. You can, however, manipulate three factors to custom program this circuit to your own needs: time base period (R1C1), the output pin on IC1 used to trigger IC2, and the strapping configuration on IC2. In other words:

$$T_o = R1C1 \times T_o' \times T_o''$$

Where T_o is total period in seconds that the output is LOW, T_o' is the period that the selected output (s) of IC1 is LOW, and the T_o' is the period that the selected output (s) of IC2 is LOW.

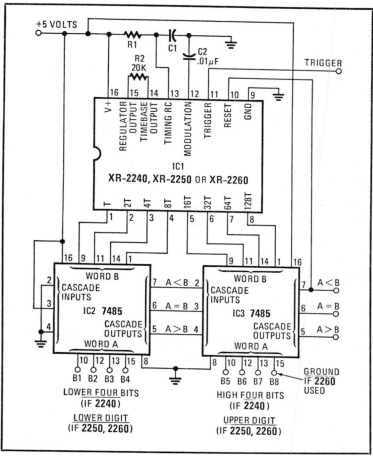

Fig. 12-19. Long-duration timer with preset count.

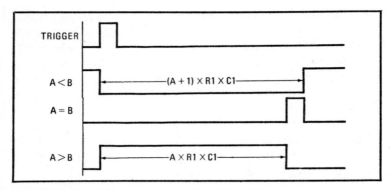

Fig. 12-20. Timing diagram of a long-duration timer.

For example, assume that the product R1C1 is one second, as if R1 = 1 megohm and C1 = 1 μF). If the output remains LOW for a period of 65536T = 65536 (R1C1) = T_o, it will remain LOW for 65536 seconds, or about 18 hours. This is on a one-second pulse! Of course, it is impossible to accurately generate a time delay this long using any of the other techniques. It is, for example, relatively common to find high-value electrolytic capacitors necessary in long-duration RC timers that are rated with a −20 percent to +100 percent tolerance in capacitance. This is incompatible with the goal of making a precision time base of long duration. In addition, most electrolytics—and that includes tantanlum—will change value over time while in service. For most filtering and bypassing applications, this is acceptable; in timing, though, it is deadly. The use of cascaded XR-2240 devides allows us to select more easily managed values of resistance and capacitance.

PROGRAMMABLE TIMERS

By adding some external circuitry, the 2240, 2250, and 2260 timers can be made programmable; i.e., an operator can select the time duration at will. Figure 12-19 shows a circuit that allows timer control with a digital word (binary for the 2240 and BCD for the 2250 and 2260). In this circuit, the output terminals are connected to a pair of 7485 TTL magnitude comparators.

The 7485 device is composed of a set of TTL exclusive-OR gates connected to compare two four-bit words and issue an output that indicates that word A is greater than B, A equals B, or that A is less than B. In this case, the output of the 2240/2250/2260 can be compared with the word applied by a microcomputer output port or an external binary or BCD thumbwheel switch.

The timing diagram for Fig. 12-19 is shown in Fig. 12-20. Because word A programs the timer, the output durations are given in terms of the value of A, and the R1C1 time base period.

When a trigger pulse is received, word A has previously been set to some value between 00000000_2 and 11111111_2. All this time, a word A is either greater than word B or equal to it if word A is also 0000000_2. In the former case, the A-greater-than-B output from IC3 is HIGH and the other two outputs are LOW. When the counter has incremented so that word A and word B are equal, then the A-equals-B output goes HIGH for a period of one clock pulse, after which word B is greater than word A, so the A-less-than-B output goes HIGH. The timing durations are (A+1) R1C1 for A-less-than-B, and AR1C1 for A-greater-than-B. The A=B output produces a single pulse at time AR1C1.

Assume that the circuit shown in Fig. 12-19 is programmed so that word A is 178_{10} (10110010_2y) and that the RC time constant is five seconds. What is the duration of the A-greater-than-B pulse? The duration is A × R1 × C1, or (178_{10}) × (5_{10} seconds), or 890 seconds (about 14.8 minutes).

Thumbwheel switches can be used with binary, BCD, or octal output encoding to program these IC timers. Figure 12-21 shows an XR-2250 timer connected to a BCD thumbwheel switch. The BCD output lines of the switch are connected to the timer output lines, and the switches are, in turn, connected to pull-up resistor R1. The switches provide a simple way to build circuits such as Fig. 12-15 using convenient front panel switches to change the output duration as desired.

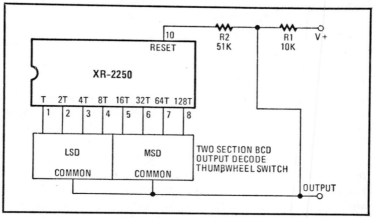

Fig. 12-21. Presettable one-shot.

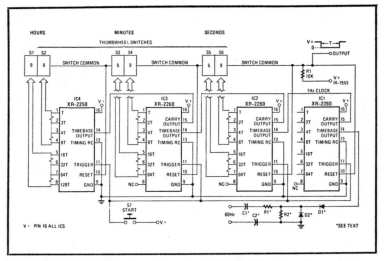

Fig. 12-22. Presettable counter in units of time (hours, sec).

Figure 12-22 shows the circuit of a precision 100-hour timer using the Exar XR-2250 and XR-2260 (or Intersil 8250/8260) timer ICs. Connected in cascade, four timers are programmed by thumbwheel switches.

Timer IC1 is used as a time base to generate 1-Hz clock pulses from the 60-Hz AC power line. The values of C1, C2, R1, and R2 depend upon the amplitude of the 60-Hz line used for the time base.

The XR-2260 counts to 59. On the 60th count, it generates a carry pulse from pin No. 15. This output pulse represents a freuqency division of 60 for the XR-2260 and 100 for the XR-2250.

As shown in Fig. 12-22, the *second* and *minutes* sections of the circuit are thumbwheel-programmed, cascaded XR-2260 ICs, while the *hours* section is a thumbwheel-programmed XR-2250 IC, allowing a maximum of 99 hours. Both XR-2260 ICs use the carry output pulse to drive the time base input of the following stage.

The thumbwheel switches program the total duration of the output pulse, but some applications might require a continuous monitoring of the time that has *elapsed* since the trigger pulse started the counters. To do this, you can use an events counter or a clock IC that uses a 1-Hz input. Another alternative is to use an hours-minutes-seconds counter circuit with its reset terminal connected to the timer output. The timer output is active-low, so it will allow counting during its time duration but will reset the counter when it goes HIGH at the end of the time duration.

Chapter 13
Data Multiplexers and Selectors

The fastest way to transfer data within digital systems or between two different digital systems is to connect all of the like-bits in parallel—all BØ lines in parallel, all B1 lines in parallel, etc. But parallel data transfers are expensive to implement. They become terribly expensive in hardware costs when there is more than one data source. If there are, say, three channels of eight bits each, 24 wires or radio communications channels are necessary to transfer the data. While this may not be too much of a problem inside the equipment cabinet, it becomes one when renting cross-city or intercity telephone lines—the standard lines used for data communications. One seemingly standard 12-bit computer is fitted with eight I/O ports. To connect them in parallel another computer or other devices, 96 wires would be needed. Try paying for 96 cross-country telephone lines.

It might not seem like much. After all, any technician worth a lot load of salt can easily and correctly connect 96 wires in a very short order. But these wires must be viewed in a little different light once outside the same room. Digital signals cannot be passed along a piece of wire longer than a few feet unless certain conditions are met. The problem is the sharp rise time of the pulses used in digital electronics. Low-impedance line drivers and line receivers must be used at the input and output ends of the line, respectively. These "wires" must be viewed as *communications channels* instead of mere wires. When you go outside your own building and have to rely on the services of the telephone common carriers, then they are indeed communications channels. How would you like to pay for 96 radio or wire communications channels? The cost is simply too much for this type of data communications.

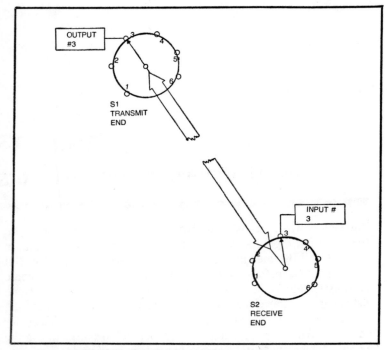

Fig. 13-1. Electrical circuit model of time domain multiplexers.

The solution is in *serial data transmission*, the subject of the next chapter. In this chapter, though, we will cover how to make several data channels fit into one communications channel, or how to make parallel data bits into serial data bits. The answer is in *multiplexing*.

MULTIPLEXING

Multiplexing is the art of *interleaving* data so that they can be transmitted together over a single line or through a single data communications channel. Two basic forms of multiplexing exist: *frequency domain* and *time domain*. The most common examples of frequency domain are color television, FM stereo broadcasting, and subcarrier telephony. All of these use one or more subcarriers to transmit information. Digital systems, however, greatly use time domain multiplexing.

Figure 13-1 shows a simplified example of a time domain multiplexer circuit. Here two rotary switches select from any of six different data channels. These switches are rotated *synchronously* so that both are always in the same position; i.e., S1 is connected to channel No. 3 at the same time that S2 is connected to channel No. 3. This situation allows us to transmit data from six channels serially along the communications path by using techniques that will be discussed more fully in the next section.

Figure 13-2 shows a system in which four eight-bit devices are connected to a single eight-bit input port. If they were connected in parallel, 4×8, or 32, wires would be necessary. Because the system is being multiplexed, however, only one I/O port (instead of four I/O ports) and 10 wires are needed. Eight of the wires are used to form the eight-bit data bus over which all of the data is to pass. The remaining two wires are used for control of the multiplexer. In the circuit shown, the multiplexer can be viewed as a programmable switch. These two wires do the programming, according to the rules shown in Table 13-1. If the code is 00, then

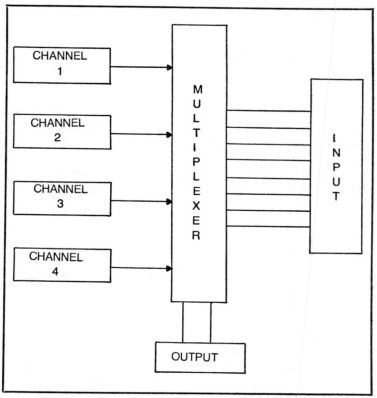

Fig. 13-2. Multiplexed system of four channels.

175

Table 13-1. Truth Table of a Multiplexed System.

Channel	Code
0	00
1	01
2	10
3	11

channel 0 is selected and connected to the data bus. If the code is 01, then channel 1 is selected, and so on. A simple two-bit binary counter, or a counter program if a computer is used, will create the codes. It will sequence 00-01-10-11-000 . . . until turned off or disabled.

The circuit shown in Fig. 13-2 assumes that the data at the inputs will always be ready, because it operates in a quasi-asynchronous mode. Synchronous operation would require at least one more wire. Of course, it must also be assumed that the device receiving the data will not change the data selector code until it has successfully completed the current data transfer. It will not, for example, switch to channel No. 2 until it has succesfully input the data on channel No. 1.

Figure 13-3 shows another form of digital multiplexing, used extensively on calculators and other portable digital instruments. LED and incandescent seven-segment display readouts (see Chapter 10) draw a considerable amount of current from the power supply. Some devices draw up to 40 mA per segment. In the four-digit display shown in Fig. 13-3, when all segments are turned on (the displayed value is 8888), the current drawn is:

$$\frac{0.04 \text{ amperes}}{\text{segment}} \times \frac{8 \text{ segments}}{\text{digit}} \times 4 \text{ digits} = I$$

$$0.04 \times 8 \times 4 \text{ amperes} = 1.28 \text{ amperes}$$

If this were a 12-digit portable calculator, even more current capability (almost 4 amperes) would be needed. Once you begin to require so much current on a continuous basis, battery life drops rapidly. If a battery large enough to allow lengthy operation is used, then portability becomes questionable, at the least.

The answer to the battery life dilemma is to multiplex the digits. Only one digit at a time will be turned on, but if the switching rate is fast enough, the *persistence* of the human eye will make the display *appear* constant.

Figure 13-3 shows a multiplexed digital display, while **Fig.** 13-4 shows the timing diagram. Recall from Chapter 10 that a seven-segment readout requires one wire for each segment; these are usually labeled a,b,c,d,e,f, and g. In a typical device, power is applied to the anode of the digit, and the decoder will ground the terminals to those segments that are to be lighted. To display a 3, for example, ground terminals a,b,c,d, and g (see the inset of Fig. 13-3). The instrument will place the proper seven-segment code for the digit to be displayed on the seven-bit data bus at the appropriate time. Timing is critical to the proper operation of this system, and it is handled by the control logic section.

The *digit selector* is a circuit that enables each digit in sequence. In the crude example shown here, a digit is selected by applying +5 VDC to the common anode. This circuit has four output lines, one for each digit. When a selector line is HIGH, then +5V is applied to its digit.

In the situation shown in Fig. 13-3, the four-digit number 4137 is displayed. The sequential operation is as follows:

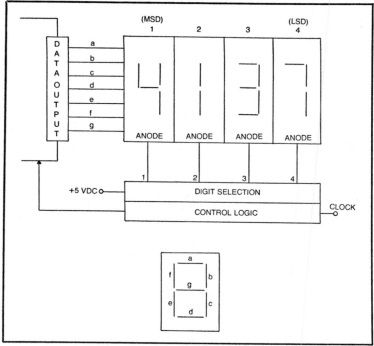

Fig. 13-3. Multiplexed seven-segment display.

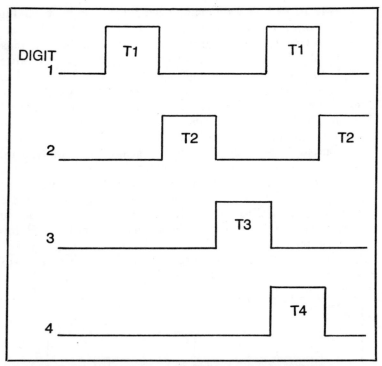

Fig. 13-4. Timing diagram of the multiplexed seven-segment display.

☐ During time T1, the seven-segment code that represents "4" is placed on the data bus. Simultaneously, the digit selector applies +5 VDC to digit No. 1. This causes digit No. 1 to display a "4."

☐ During time T2, the seven-segment code representing "1" is placed on the data bus. Simultaneously, the digit selector applies +5 VDC to digit No. 2 after having disconnected all other digits.

☐ During time T3, the seven-segment code representing "3" is placed on the data bus. Simultaneously, the digit selector applies +5 VDC to digit No. 3, which turns off the other three digits.

☐ During time T4, the seven-segment code representing "7" is placed on the data bus, and the digit selector turns on digit No. 4.

This sequence happens continuously. Only one digit is turned on at any one time, but they all turn on sequentially many times per second.

A simple example of a two-channel multiplexer is shown in Fig. 13-5. This circuit uses a common CMOS digital integrated circuit, the 4016, or its newer and preferred cousin, the 4066. Both 4016 and 4066 devices are easily available through mail order electronics houses specializing in IC sales. Note that the CMOS electronic switches used in these devices are *bilateral*, meaning that current will pass in either direction. This means that they can be used as either analog, where bipolar signals might be encountered, or digital multiplexers.

Each of the switches inside the 4016/4066 devices is completely independent of the others. They share only common power-supply terminals.

The control pins are 5, 6, 12, and 13. When the voltage applied to these pins is equal to the voltage on pin No. 7, the switch selected by that particular pin is turned off. In digital systems, pin No. 7 will most likely be grounded, so the switches are turned off when their control pins are grounded (zero volts). The voltage applied to pin No. 7 can be anything in the range 0V to −5V (−7V in some models), but the negative voltage are only seen occasionally in digital circuits.

Applying a voltage equal to the voltage on pin No. 14 to a control pin will turn *on* the switch. In most digital circuits, this will be +5V, although CMOS devices can operate up to +18 VDC. The rationale is to make the whole circuit TTL compatible.

The terms "on" and "off" refer to the series impedance of the switch. In the off condition, the series impedance is very high,

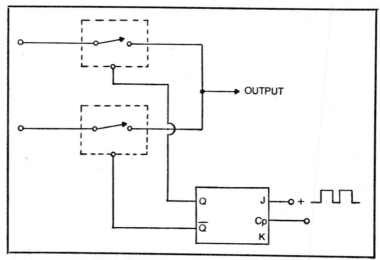

Fig. 13-5. Two-channel multiplexer based on 4016/4066 device.

usually well over a megohm. In the on condition, however, the series impedance of the switches drops very low. It will be in all cases less than 2000 ohms, and in some devices less than 100 ohms.

In the "muxer" of Fig. 13-5, only two of the switches are used. This circuit, incidentally, is also useful for making a single trace oscilloscope into a dual trace scope. Note that one terminal of both S1 and S2 are connected together to form a common output. The remaining terminal on both switches goes to the respective channels. Pin No. 1 is connected to channel No. 1, and pin No. 4 is connected to the channel No. 2.

The switching action is performed by a J-K flip-flop connected in the clocked binary divider method. The Q and not-Q outputs on a J-K flip-flop are complementary. This means that one output will be HIGH when the other is LOW. When the state of the FF changes, the HIGH output becomes LOW and the LOW output becomes HIGH. The timing diagram shown in Fig. 13-5 shows this relationship more clearly.

During interval T1, the Q output is HIGH and not-Q is LOW. This turns on switch S1 and turns off S2 (pin No. 5 is HIGH and pin No. 13 is LOW). According to the rules for operation of the 4016/4066, channel No. 1 is then connected to the output.

Following the next clock pulse (J-K flip-flops operate on the negative-going edge of the clock pulse), the condition of the outputs becomes reversed. S1 is now turned off and S2 is turned on. This will connect channel No. 2 to the output.

The switches are turned on and off alternately. This means that the output will be connected to the respective inputs alternatively.

In the case of an oscilloscope switch, only this one circuit is needed to make things happen. In data communications systems, however, the signals are now interleaved, and that makes things a hopeless nightmare at the other end of the transmission path unless some form of decoding is provided. A *demultiplexer* must then be provided at the receive end of the path.

Happily, the 4016/4066 devices are bilateral. Basically the same circuit is useful for "demuxing." The serial data stream coming into the receiver is fed to the common terminal between the two switches (used as an output in the mux end and an input in the demux end).

Three additional standard CMOS devices are useful as multiplexers and demultiplexers: 4051, 4052, and 4053. All three of

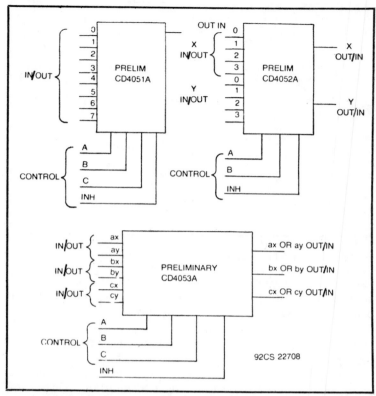

Fig. 13-6. 4051, 4052 and 4053 analog/digital muxers.

these CMOS devices have chip enable pins that allow the chip to be turned on and off at will.

The 4051 shown in Fig. 13-6 is a 1-of-8 selector and can be viewed as analogous to a single-pole, eight-position rotary switch. Again, like the 4016/4066 devices, the switches are bilateral. Three control terminals (A/B/C) select which one of the switches is turned on at any given time. These are programmed according to the ordinary binary sequence for the digits 0 through 7, such as 000 on A/B/C causes channel 0 to be on, 001 causes channel 1 to turn on, and so on. A binary counter, driven by a clock, will sequence through all eight possible conditions.

The 4052 device shown in Fig. 13-6 is a little different. It contains two independent 1-of-4 switches. The independence is not total, however. There is only one set of control terminals: A/B. These respond to the codes for 0 through 3. When A/B sees 000,

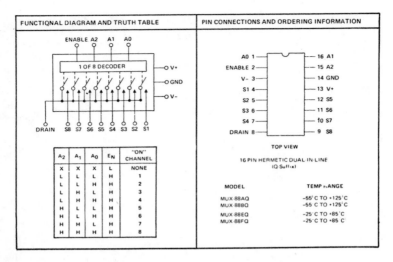

| | FUNCTIONAL DIAGRAM AND TRUTH TABLE | PIN CONNECTIONS AND ORDERING INFORMATION |

A_2	A_1	A_0	E_N	"ON" CHANNEL
X	X	X	L	NONE
L	L	L	H	1
L	L	H	H	2
L	H	L	H	3
L	H	H	H	4
H	L	L	H	5
H	L	H	H	6
H	H	L	H	7
H	H	H	H	8

TOP VIEW

16 PIN HERMETIC DUAL-IN-LINE
(Q Suffix)

MODEL	TEMP RANGE
MUX-88AQ	–55°C TO +125°C
MUX-88BQ	–55°C TO +125°C
MUX-88EQ	–25°C TO +85°C
MUX-88FQ	–25°C TO +85°C

Fig. 13-7. Precision Monolithics MUX-88.

both channel 0s are turned on. When the code is 01, both channel 1s are turned on. This process continues. We can view the 4052 as a pair of single-pole, four-position rotary switches.

The 4053 device shown in Fig. 13-6 is a triple 1-of-2 switch. It is, therefore, analogous to a three-pole, double-throw switch, or, perhaps, three SPDT switches ganged together.

All three of these CMOS devices can be used as multiplexers and demultiplexers. In most cases, though, the 4051 would be used for a single channel. If you were to multiplex together more than one data source, the other devices could be used.

IC MULTIPLEXERS/DEMULTIPLEXERS

The previous examples have been regular 4000-series CMOS devices pressed into service as mux/demux devices. The trend in recent years had been to use specialized integrated circuits that employ FET technology for the muxer/demuxer. In this section, certain commercial examples that have become popular are covered.

Figure 13-7 shows the Precision Monolithics, Inc. (PMI) MUX-88 multiplexer/demultiplexer device. It is a monolithic eight-channel switch using *junction* field-effect transistors, instead of the MOSFETs used in CMOS devices. This substantially reduces the susceptibility of the device to damage by static electricity discharges.

182

The MUX-88 is versatile enough to operate from either TTL or CMOS logic levels and can therefore be used in almost any existing application. The switching action is make-before-break.

As in the simple example of Fig. 13-5, the output of the MUX-88 is connected to one side of all eight switches. The remaining terminal of each switch will go to the respective channels. Like the 4051, this device responds to a three-bit (octal) binary code that determines which switch is to be turned on at any given time. There is also a chip enable terminal that turns on and off the MUX-88.

The Datel MX-series of multiplexer/demultiplexers is shown in Fig. 13-8. This family of IC devices is useful in a variety of applications because it is compatible with TTL, CMOS, and even the old DTL logic system. Like the MUX-88, these are monolithic integrated circuits, rather than discreet or hybrid technology devices. There are different sizes of multiplexers in this family, but

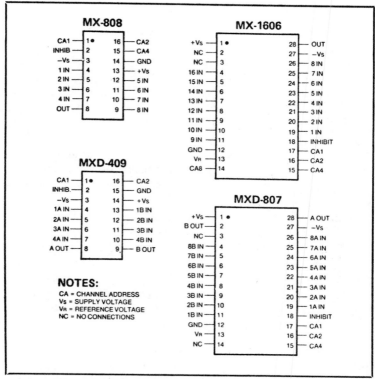

Fig. 13-8. Pinouts for MX-series muxers.

8	4	2	1	INHIB.	ON CHANNEL
X	X	X	X	0	NONE
0	0	0	0	1	1
0	0	0	1	1	2
0	0	1	0	1	3
0	0	1	1	1	4
0	1	0	0	1	5
0	1	0	1	1	6
0	1	1	0	1	7
0	1	1	1	1	8
1	0	0	0	1	9
1	0	0	1	1	10
1	0	1	0	1	11
1	0	1	1	1	12
1	1	0	0	1	13
1	1	0	1	1	14
1	1	1	0	1	15
1	1	1	1	1	16

Table 13-2. MX-1606 Channel Addressing.

all use either 2-, 3-, or 4-bit binary codes as the switch select signals. Tables 13-2 through 13-4 show channel addressing.

One advantage the specialized CMOS devices has over standard devices is that transfer accuracy is generaly guaranteed more closely. In the Datal MX-series, for example, 0.01 percent transfer accuracy is obtainable at sample rates up to 200 kHz. The devices will tolerate bipolar signal excursions over the range ±10V, so they can be used for either analog or digital muxers.

This series requires a power supply in the range ±5 to ±20 VDC. Interestingly, the power consumption is only 7.5 mW (standby condition). Even at a high sampling rate, in the 100-kHz range, the power consumption is only 15mW.

Be careful about the total power *consumption* and the total package dissipation. The former refers to the amount of power required from the DC supply in order to power the internal circuits of the device. The total dissipation refers to the amount of power that can be dissipated by the device and includes the power loss in the switches. Because each switch has a series resistance—even though small—there will be a signal loss, and this translates as a power dissipation. In the Datel MX-series, the total allowable dissipation is on the order of 725 mW for the MX-808 and 1200 mW (that's 1.2 watts!) for the MX-1606 and MXD-807 devices.

Table 13-3. MX-808 and MX-807 Channel Addressing.

4	2	1	INHIB.	ON CHANNEL
X	X	X	0	NONE
0	0	0	1	1
0	0	1	1	2
0	1	0	1	3
0	1	1	1	4
1	0	0	1	5
1	0	1	1	6
1	1	0	1	7
1	1	1	1	8

The MX-1606 is a 1-of-16 channel single-ended multiplexer. You might view it as an electronic SP8T switch. It uses a four-bit address select code; four bits are required for 16 channels because $2^4 = 16$. Similarly, the four-channel device (MXD-409) and the eight-channel device (MXD-807) use two-bit and three-bit select codes, respectively.

The transfer accuracy of any electronic switching multiplexer depends upon the source and load resistances. The output voltage is given by the standard voltage divider equation:

$$E_{out} = \frac{E_{in} R_L}{R_s + R_{on} + R_L}$$

where E_{out} is the switch output voltage, measured across load resistor R_L; E_{in} is the input voltage, or the open-circuit voltage of the input signal source; R_L is the load resistance; R_s is the output impedance of the signal source; and R_{on} is the resistance of the switch when *on*.

It doesn't take a mathematical genius to figure out that it is wise to keep the R_s and R_{on} terms of the equation as low as possible, and to keep the R_L term as high as possible. We can meet the low-R_{on} criterion by using an operational amplifier with a low

Table 13-4. MXD-409 Channel Addressing.

2	1	INHIB.	ON CHANNEL
X	X	0	NONE
0	0	1	1
0	1	1	2
1	0	1	3
1	1	1	4

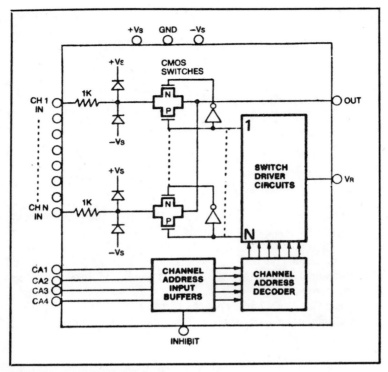

Fig. 13-9. Datel MX-series muxers.

output impedance at the input of the electronic switch. The load resistance requirement can also be met by using an operational amplifier. It should have an input impedance that is greater than 10^7 ohms. This is not too hard to obtain because some low-cost BiFET or BiMOS input op amps boast input impedances in the 10^{12} ohms range.

□ **Example.**

Find the output voltage error in percent if the following terms apply: $R_s = 100$ ohms, $R_L = 10^7$ ohms, and $R_{on} = 10^3$ ohms.

$$\frac{E_{out}}{E_{in}} = \frac{R_L}{R_s + R_{on} + R_L} = \frac{(10^7)}{(10^2) + (10^3) + (10^7)}$$

$$= 9.9989 \times 10^{-1} = 0.99989$$

Converting to percent:

$$S = \frac{1 - 0.99989}{1} \times 100\%$$

186

$$= \quad (0.00011) \, (100\%)$$
$$= \quad 0.011\%$$

In this example, you can see that an error of approximately 0.01 percent will be obtained when we use values that are reasonably close to those actually obtained in the real world. To reduce this error even further, then, use an operational amplifier with a very high input impedance. If an RCA CA3130, CA3140, or CA3160 were used with a 10^{12}-ohm input impedance, the error would reduce to a ridiculously small value—so small that it is difficult to calculate on a 10-digit electronic calculator. Devices such as the LF156 are also used for this purpose.

Figure 13-9 shows a scheme for using four 16-channel muxes, such as the Datel MX-series (MX-1606), to make a 64-channel multiplexer. A 6-bit select code is required to determine the channel that is turned on at any one time. To minimize loading, the common outputs are connected together at the noninverting input of a unity gain operational amplifier.

Chapter 14

Data Transmission

Data transmission involves moving digital data from one location to another. The transmission path might consist of only a few yards across the room or down the hall. It might also consist of a long-distance communications system involving telephone (wire) communications, radio or microwave links, long-distance coaxial cable hookups, or even satellite communications. In this chapter, we are going to discuss some of the fundamental aspects of data communications, both local (inside the same building) and long distance.

PARALLEL DATA COMMUNICATIONS

Short-distance data communications can be carried out over parallel lines. The only requirement here is to have one wire for each bit and control signal needed to complete the job.

One of the simplest examples is the ordinary 7-bit ASCII keyboard interface to a computer. Such a keyboard interface receives the binary word for an ASCII alphanumeric or machine control character on the lower seven bits (BØ through B6) of the port. The high-order bit (B7) of the port is connected to the keyboard *strobe* output. The strobe becomes active when the data on the lower seven bits correctly represents the desired character. Such a circuit is shown in Fig. 14-1. The programmer of the computer will have to write a keyboard program that looks something like Fig. 14-2. The computer constantly tests for B7 being active (let's assume active-HIGH). If B7=LOW, then the program jumps back to the beginning and starts over again; it continues to loop through, testing for B7=HIGH. But if it finds B7=HIGH, the active condition in this example, the keyboard data is valid—only

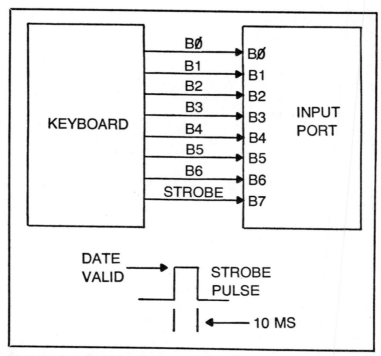

Fig. 14-1. Connecting a keyboard to an input port.

one strobe pulse is generated for each keystroke, so the program jumps to an input instruction. After the main program "does its thing" with the input data, the program jumps back to the beginning of the loop.

Figure 14-3 shows another possible situation involving parallel data transmission. In this case, two computers are connected together; it could just as easily be a peripheral and a computer—the principle is the same. Such a circuit may or may not use the control signals, depending upon whether the data transfer is to synchronous or nonsynchronous. This circuit merely requires a bit of cable, usually flat ribbon cable in computer systems, and a couple of connectors. The programmer will write a subroutine program to interface the data transfer.

In the case of nonsynchronous, or asynchronous, data transfer, a loop program such as given for the keyboard is used. In fact, the keyboard represents an example of an *asynchronous peripheral*. Sometimes, though, the computer will be doing something else when the peripheral needs servicing. Alternatively, it could be that

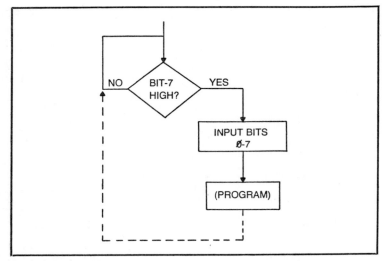

Fig. 14-2. Program flow chart.

the two devices have vastly different operating speeds. This frequently occurs in the case of electromechanical peripherals sending or receiving data from computers. The data transfer might have to be synchronized in these cases. And the signals between the two devices need to be controlled to cause synchronization. This is sometimes called *handshaking*. By sending a pulse to device B, device A will tell device B, "Hey! I'm ready to receive the next data byte." By returning another pulse, Device B will reply, "OK, here it comes." While it is usually a little more complicated than this, the idea is the same.

SERIAL DATA CIRCUITS

Serial data bits are transmitted one at a time. Perhaps the simplest example of a serial data system is the old-fashioned Morse code. The operator would tap out the dots and dashes of the code one at a time, and that is how they would be transmitted along the system or over the air. Figure 14-4 shows an example of a serial data stream. In this case, the binary number 11001001 is being transmitted. The binary digits can take on only one of two values (HIGH or LOW) at any one time. The key is in knowing exactly when a given bit is expected and then looking to see the condition at that instant. Because of this requirement, it is critical to know when the word begins and when it ends. It is also critical that the speed of transmission at the receive end be matched within a very

190

small error to the speed of transmission at the sending end. The beginning of the word will be denoted by the existence of a start bit. Ordinarily, the line will be held in one condition or the other; it then switches to the condition of the start bit when the transmission begins. The start of a word is very often indicated by a start bit that is 1.5 or 2 times the normal duration of the data bits.

Data speed is measured in bits per second, sometimes called *baud*. The precise definition of baud is somewhat more involved and subject to some differences of opinion, but bits per second is used most frequently. You will sometimes see the term baud rate, but because baud *is* a rate, it is redundant; nevertheless its use is so frequent that we must nod to standard practice and forget proper word usage.

It is necessary that the baud rate be matched at both ends of the serial data communications system. Because of this, it has become standard to use one of only a few different popular rates: 110, 300, 600, 1200, 2400, 4800, and 9600 baud. The slowest rate, 110 baud, is one of the more popular used in printers and teletypewriters. At 110 baud, it is possible to transmit approximately 10 characters per second, each character requiring 100 ms. At this

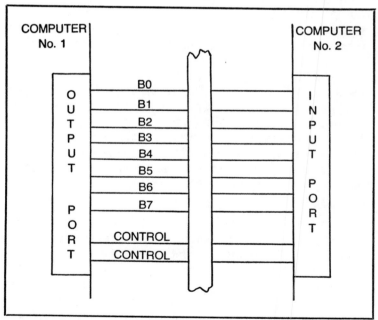

Fig. 14-3. Connecting two computers together.

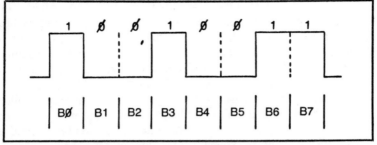

Fig. 14-4. Serial representation of binary word 11001001.

rate, teletypewriter speeds of 100 words per minute can be accommodated, with time left to spare for such housekeeping chores as carriage return and line feed. As a general rule, most devices use baud rates of 1200 or less because it is possible to use ordinary telephone circuit. Faster rates require special conditioned lines that are a lot more costly. These special lines are often dedicated to only that one data communications link, thereby increasing the cost even further if you do not need the link all of the time.

Serial data communications systems can be classified into the following: simplex, half-duplex, and full-duplex. The *simplex* system permits communications in only one direction. This is like ordinary radio broadcasting or radio control applications, where a fixed transmitter at one end of the path communicates with a fixed receiver at the other end. Half-duplex is a system in which two-way communications is possible, but not simultaneously; only one direction at a time is used. This is analogous to two-way radio systems. Operator A can talk to B, but B cannot talk back until A is finished and relinquishes the air. Full-duplex is a system in which two-way communications can exist simultaneously. In those systems, the data can flow in both directions at once. Each end of the communications system will have its own receiver and transmitter that are completely independent of each other.

INTERFACE STANDARDS

Several different serial data interface standards have been adopted by the data communications industry. Among these are RS-232, RS-422, GPIB (general-purpose interface bus), and IEEE-488. The purpose of these standards is to provide standard voltage levels, operating speeds, and connector pinouts so that equipment made by different manufacturers will be directly compatible with each other. A computer manufacturer, for example, might provide an RS-232 interface card with the machine. This means that a peripheral manufacturer can sell equipment that will interface directly with the computer, simply making the peripheral to the RS-232 standard. A complete range of modems (modulator-demodulator), CRT video terminals, printers, teletypewriters, and other peripheral devices are available in RS-232. This is probably the most popular standard of the Electronic Industries Association (EIA).

Unfortinately, the RS-232 is one of the older standards used by the industry. It predates TTL, so the logic levels of the RS-232 system are not TTL compatible. This is something of a disadvantage, but common usage apparently keeps RS-232 alive and well.

The RS-232 standard specifies the logic level voltages and the impedances so that devices of different manufacturers have some small hope of operating together with success. Figure 14-5 shows the RS-232 voltage levels. Note that there are actually two versions of the RS-232 standard. The older standard, RS-232B, has a wider spread between the two voltage levels used to represent the two logic states. logical-∅ was present when the voltage was between +5V and +25V, while logical-1 existed when the line voltage was between −5V and −25V.

On the other hand, RS-232C uses narrower ranges. The logical-∅ level was +5V to +15V, while logical-1 was −5V to −15V. This change resulted in speedier operation of the circuit:

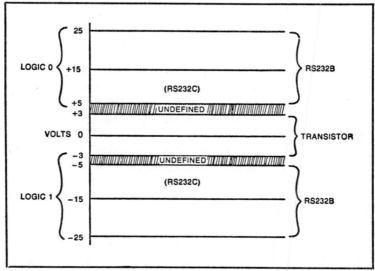

Fig. 14-5. RS-232B and C voltage levels.

Table 14-1. RS-232 Pin Definitions.

Pin No.	RS232 Name	Function
1	AA	Chassis ground
2	BA	Data from terminal
3	BB	Data received from modem
4	CA	Request to send
5	CB	Clear to send
6	CC	Data set ready
7	AB	Signal ground
8	CF	Carrier detection
9	undef	
10	undef	
11	undef	
12	undef	
13	undef	
14	undef	
15	DB	Transmitted bit clock, internal
16	undef	
17	DD	Received bit clock
18	undef	
19	undef	
20	CD	Data terminal ready
21	undef	
22	CE	Ring indicator
23	undef	
24	DA	Transmitted bit clock, external
25	undef	

Under worst case conditions, it takes less time to make the transition from −15V to +15V than it does from −25V to + 25V.

Designers of RS-232C circuits must contend with ±25V logic levels, driver output impedance specs, receiver input impedance specs, and a 30/Vus slew-rate requirement. Unfortunately, TTL devices do not meet the logic level specs of RS-232C, yet they make up the bulk of the digital devices connected to RS-232C circuits. In most cases, designers use one of the ready-made IC level translators designed especially for RS-232C. The Motorola MC1488 line driver and MC1489 line receiver are examples.

The connector used for RS-232C is a standard type-D connector, commonly used for many different computer applications. This is a 25-pin connector that is readily available. It is sometimes called a DB-25 connector. At other times it is known as an RS232 connector. The latter is a misnomer that is commonly used in the digital industry. Table 14-1 shows the pinouts for the RS-232 25-pin standard connector.

TELETYPEWRITER CIRCUITS

Teletypewriters are electromechanical typewriters that will type under automatic control. Basically, such machines use a mechanism similar to an electric typewriter. Instead, though, a teletypewriter has a series of five or seven solenoids to pull in the selector bars of the typewriter for a desired character. The original teletypewriters were made by the Teletype Corporation of Skokie, IL. Note that *Teletype* is a trademark of the Teletype Corporation.

These machines operate using a series current loop for the electromagnets that pull in the selector bars. Figure 14-6 shows a recommended interface circuit that allows computer control of the current loop. One bit (usually LSB) of a computer output port is used to control the circuit. It is applied to the base of switching transistor Q1. The user must provide a 130 VDC power supply and a rheostat (R2). With the circuit connected Q1 turned on, and a key on the teletype pressed to close the loop, R2 is adjusted for a current of approximately 60 mA.

Bit control is gained by turning Q1 on and off. When Q1 is turned on, the loop is closed, and current will flow. This is the logical-1 condition and is called a *mark* in teletypewriter terminology. Alternatively, when the input bit to Q1 is LOW, Q1 is turned off. The loop is now open, and no current will flow. This is the logical-Ø condition, called a *space*. Diode D1 is used as a transient suppressor. The electromagnets of the teletypewriter will produce a sharp spike when the loop is opened due to inductive kick.

The simple circuit of Fig. 14-6 is too simple, and there are some unfortunate problems with it. One is the fact that the TTL

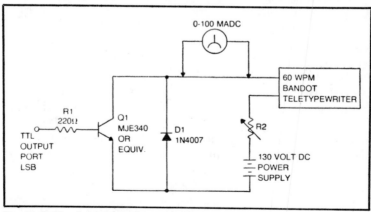

Fig. 14-6. 60 mA teletypewriter current loop connection.

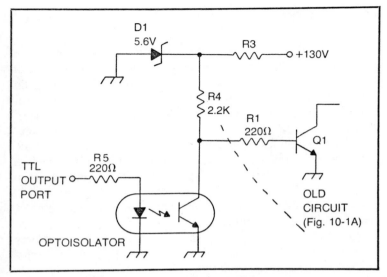

Fig. 14-7. Isolated version of a current loop connection.

output port of the computer is not entirely safe if Q1 develops a collector-to-base leakage current of any significant magnitude. The 130 VDC supply has been known to damage the output port under that condition. The second problem is that spikes from the circuit, those not entirely suppressed by D1, will get into the computer because the circuit is not well isolated. The modification of the circuit shown in Fig. 14-7 solves both of these problems. In this case, the teletypewriter loop control transistor is turned on and off by the LSB of the computer output port, but through an optoisolator. Such a device consists of a light-emitting diode (LED) and a phototransistor inside a common package. When the diode is dark, the transistor is not biased and will therefore not conduct any C-E current. When the LED is turned on, however, the base of the transistor is forward biased, so the transistor is turned on. A HIGH on the LSB of the computer output port will turn on the LED, causing the optotransistor to turn on. This will short out the bias to the loop control transistor (Q1), making it see a LOW. Of course, this circuit requires us to use inverted code from the computer, but that is no problem. We could also add one additional common emitter stage or even a TTL inverter to rectify the problem.

The circuit in Fig. 14-6 works only with the Baudot encoded, 60 wpm teletypewriters. These machines (Model 15, 21, etc.) are old and no longer currently sold. Of course, older systems—and

196

amateur radio operators using RTTY and amateur computerists—will still use such machines.

Most modern systems will use a more recent machine. One of the standards of the industry is the Teletype Model 33. This machine uses a 20 mA (instead of 60 mA) current loop. It is encoded in ASCII, which is more compatible with computer practices. The model 33 is capable of 100 wpm (110 baud) operation. A circuit for interfacing the model 33 is shown in Fig. 14-8. The terminal strip shown is found on the rear of the teletypewriter, underneath the removable top cover, on the right-hand side as viewed from the front of the machine. It is necessary that you identify the pin numbers, especially plus pins 1 and 2 are carrying 110 VAC. These can be recognized as having white and black wires, respectively. This circuit does not use the internal 20-mA power supply; instead, it uses external power supplies. For proper isolation, however, it is necessary to use ±5V and −12V power sources separate from the computer power supplies. For even greater isolation, use separate power supplies *and* an optoisolator (Fig. 14-9). In Fig. 14-9 the transistor in the previous circuit is replaced with the transistor to add optiosolator protection to the *send* side of the circuit.

Later models of Teletype, such as the Model 43, are made more computer compatible. These are available in options that

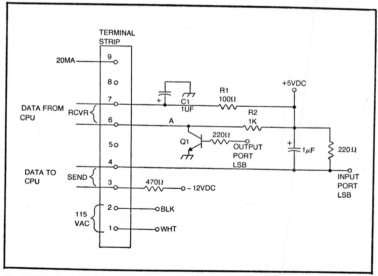

Fig. 14-8. Connection to the control terminal block of a model 33 Teletype®.

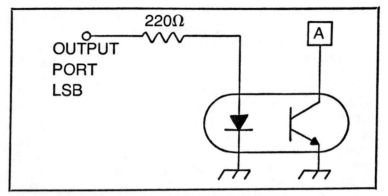

Fig. 14-9. Modification of the model 33 for better isolation.

permit either 20-mA current loop operation or RS-232, ASCII operation.

UNIVERSAL SYNCHRONOUS RECEIVER-TRANSMITTER CHIPS

The design of digital circuits for serial data transmission was once something of a chore. It required a complex selection of gates, registers, and other logic elements. Today, it's different. Some semiconductor manufacturers offer single LSI chips that will do the entire job called universal asynchronous receiver-transmitter (UART) chips.

Asynchronous transmission is preferred over synchronous transmission because it is not necessary to precisely track the clocks at each end. The clocks must be operating at very nearly the same frequency, but they need not be locked together. This eliminates the added circuit or extra communications channel needed to synchronize the two clocks. The tolerance of the clock rates in asynchronous transmission is said to be 0.01 percent, but this is easily obtainable with not too much trouble.

The UART is a single large-scale integration chip that will perform all of the data transmission functions on the digital side. It is programmable as to bit length, parity bits, and the overall length of the stop bits.

The block diagram to a common UART IC is shown in Fig. 14-11. This particular device is the TR1602A/B by Western Digital. Note that the pinouts for the UART are almost universally standardized and are based on the now obsolete AY-1013. Most UARTs are capable of all three communications modes: simplex, half-duplex and full-duplex. This feature is due to the fact that all of the transmitter and receiver control pins are independent of each other.

The UART is capable of being user-programmed as to transmitted word length, baud rate, parity type (odd-even, receiver verification/transmitter generation), parity inhibit, and stop bit length (1, 1.5 or 2 clock periods). The UART also provides six different status flags: transmission completed, buffer register transfer completed, received data available, partity error, framing error, and overrun error.

The maximum clock speed is between 320 kHz and 800 kHz, depending upon the particular type selected. Note that the clock rate actually used in any given application is dependent upon the baud rate. The clock frequency is 16 times the data baud rate.

The receiver output lines are tri-state, which means there is a high impedance to both ground and V+ when the outputs are inactive. This allows the outputs of the receiver section to be connected directly to a data bus without extra circuitry.

The transmitter section uses an eight-bit input register. This makes it capable of accepting an eight-bit parallel word from a source such as a keyboard, computer output port, etc. It will assemble these bits and then transmit them at the designated time, adding any demanded parity or stop bits.

The receiver data format is a logical mirror image of the transmitter section. It will input serial data bits, strip off the start, stop and parity bits (if used), and then assemble the binary word in parallel form. In addition, it will test the data for validity by comparison of the parity bits and stop bits.

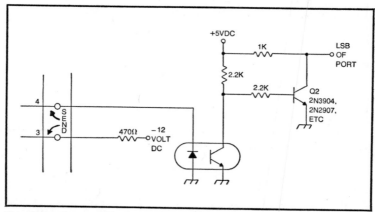

Fig. 14-10. Transmit connections for the model 33.

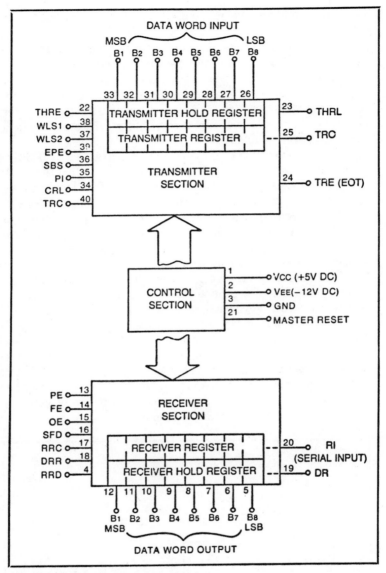

Fig. 14-11. Block diagram of a UART.

The standard UART data format is shown in Fig. 14-12. The data line (transmitter serial output or receiver serial input) will normally sit at a logical HIGH level unless data is being transmitted or received. The start bit (BØ) is always LOW, so the HIGH to

LOW transition is what the UART senses for the starting of a word or transmission. Bits B1 through B8 are the data bits loaded into the transmitter register from the outside world. Although Fig. 14-12 shows all eight bits, the user can program the device for fewer if needed. Bit lengths of 5, 6, 7, or 8 bits are allowed. Bits are dropped in shorter formats from the B1 side of the chain.

Figure 14-13 shows a typical receiver and transmitter configurations for the UART. The transmitter section is shown in Fig. 17-11A, while Fig. 17-11B shows the circuitry for a receiver section. If the serial output of the transmitter is connected to the serial input of the receiver, a closed loop exists. The output word from the receiver would match the transmitted word. In most cases, the UART is used to drive some external communications channel, such as an audio-frequency shift keyer, for transmission over some standard communications media.

Neither digital levels representing the two logic states nor binary bits can be applied directly to telephone lines or two radio communications channels. Most common radio and telephone channels want tones in the 300 to 3000 Hz range with some extended bandwidth facilities wanting to see 100 to 10,000 Hz. Before bits can be transmitted along one of these media, they first must be converted to audio tones that are compatible with the telephone lines or radio modulator input. Audio-freuqency shift keying (AFSK), similar to that used in some forms of teletypewriter communications, is generally used to represent the two tones.

Various standards have been developed that assign tone pairs (one each for logical-1 and logical-0) that may be used. Again, this is done for the sake of standardization throughout the industry. In some cases, manufacturers will use the Bell System's *Dataset*® tones; in others, they will use one of the other actual or de facto standards. Some use the regular telephone *Touch-tone*® frequen-

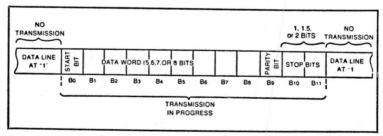

Fig. 14-12. Serial data format from UART.

Table 14-2. Touch-Tone Frequencies.

Digit	Tone	Frequencies
0	941 and	1336 Hz
1	697	1209
2	697	1336
3	697	1477
4	770	1209
5	770	1336
6	770	1477
7	852	1209
8	852	1336
9	852	1477
*	941	1209
#	941	1477
	(Low group)	(High group)

cies, in which *two* frequencies are used to represent each of 10 digits plus several control signals. Each character is represented by a high frequency and a low frequency. Table 14-2 lists the several Touch-tone frequencies.

Touch-tone signals are most easily generated using a specialized integrated circuit, such as the Motorola MC14410. This chip is an LSI CMOS device that is designed to generate the two frequencies required on command from a standard two-of-eight contact closure keypad. Several manufacturers sell these pads in a standard format. Tpey are arranged in a 4×4 matrix: four rows and four columns. When a pushbutton is depressed, a crossbar switch shorts together one row and one column, creating a unique logic combination out of a possible 16.

The MC14410 contains its own timing circuits, including a clock oscillator that operates off a standard 1000-kHz crystal (external). The chip creates two semi-sine waves, digitally using a special addition technique. The tolerance of the output frequency is on the order of 0.2 percent, well within the Touch-tone specification.

The basic connection scheme for the MC14410 is shown in Fig. 14-14. Note that very little in the way of external circuitry is required on this chip; it is almost self-contained. The only connections required in this most basic configuration are four row inputs, four column inputs, ground, +5 VDC, and the high-frequency and low-frequency output terminals. Note that the two outputs are independent of each other, allowing them to be used separately, if needed. In most applications, however—especially actual tele-

202

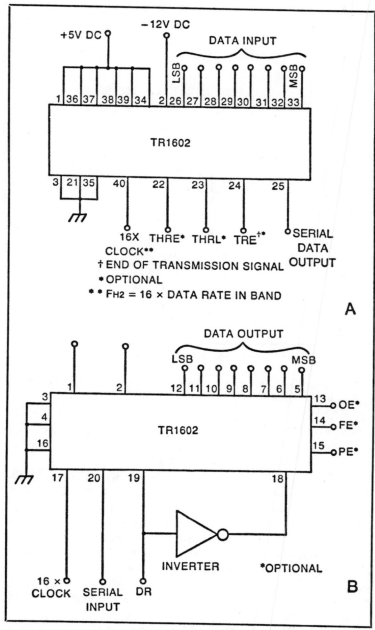

Fig. 14-13. Transmitter connections on UART at A, and receiver connections on UART at B.

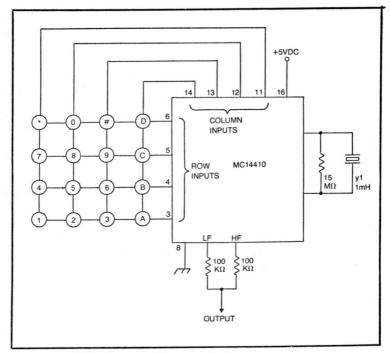

Fig. 14-14. Connection of the MC14410 Touch-tone generator.

phone circuits—these outputs are summed in a resistor mixing network.

The keyboard inputs to the MC14410 want to see a LOW to ground in order to be selected. The keyboard will simultaneously, in crosspoint manner, enable only the two inputs that correspond to the number or character on that key top.

An example of a frequency shift keyer is shown in Fig. 14-15. In this circuit, the two frequencies (F1 and F2) are generated in clock oscillator circuits or in a countdown circuit from a master clock (as appropriate). Two TTL NAND gates are used to control the output frequency. F1 is applied to one input of G1, while F2 is applied to one input of gate G2. The data line is the binary bit that is being represented; i.e., it is HIGH for a logical-1 and LOW for a logical-Ø. When the data line is LOW, then gate G2 is inhibited; according to the rules for a NAND gate, it will have a HIGH output if either input is LOW. Because of the inverter, however, the input of gate G1 is HIGH. Its output line, therefore, is allowed to follow the F1 clock. Similarly, when the data input is HIGH, gate G1 is inhibited, and G2 follows the F2 clock. Clearly, then, F1 is output when the data line is LOW, and F2 is output when the data line is HIGH.

Figure 14-15 shows a resistor mixing network. This will sum the two outputs and reduce their amplitude, so that they can be applied to the same communications circuit. Most communications equipment that requires audio input signals will not tolerate TTL level signals, so the amplitude must be reduced. If the communications channel specifically asks to see a TTL logic level at the input, though, then replace the resistor network with a two-input NOR gate (one input tied to the G1 output and the other to the G2 output).

Several other circuits are common used, including a few special-purpose IC devices. Rarely will you find something like a voltage-controlled oscillator or an oscillator that is started up when the appropriate logic level is applied. This is too messy and tends to pull the oscillator a little when first turned on. The resultant output is not unlike chirp in CW radio transmitters.

Although circuits differ somewhat from one manufacturer to another, there are two basic forms of tone decoder circuits: filters and phase-locked loops (PLL). An example of the *filter* style of

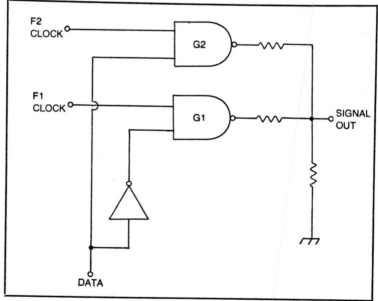

Fig. 14-15. Tone encoder for binary data.

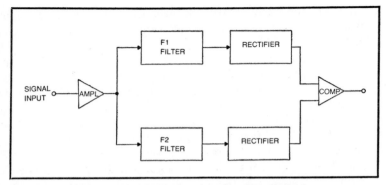

Fig. 14-16. Block diagram of tone decoder using filter method.

decoder circuit is shown in Fig. 14-16. An input amplifier may or may not be used, depending upon the circumstances. If the signal is weak and noisy, as it might well be after transmission over long-distance telephone lines, the amplifier is needed. The bandwidth of the amplifier is narrow enough to pass only the two frequencies or any additional frequencies used in control applications. This eliminates much of the noise that comes in with the signal. The output of the amplifier is applied to the inputs of two very narrow-band, high-Q, band-pass filters. One filter will pass only F1, while the other will pass only F2. The outputs, of the two filters are usually then rectified, and the rectified—but not smoothed—signal is applied to one input of a voltage comparator. Recall that a comparator circuit will produce an output that indicates whether or not the input voltages are in agreement. If F1 is present, for example, the comparator would be designed to produce a LOW output, but if F2 is seen, then a HIGH is output. In general, the comparator method is probably the most commonly used in low-speed data communications applications.

But the phase-locked loop of Fig. 14-7 is not without supporters. A phase-locked loop is a circuit that will adjust the frequency of a VCO to keep it in sync with an input signal. In a tone decoder, the input signal is the desired tone. In the circuit of Fig. 14-17, the output (pin No. 8) goes LOW when the desired tone is applied to the input. We can use two of these circuits—one for each tone—to decode AFSK signals used in data communications with a little extra logic circuitry. For the most part, however, this type of circuit is used in control applications where the tone controls a computer, other processor, or radio repeater. Again by the use of appropriate external logic circuits, the entire range of 16 possible

206

input signals can be decoded into 1-of-16 outputs. The response frequency of this circuit is given approximately by:

$$F = \frac{1}{(R1 + R2)\, C1}$$

where F is the frequency in hertz, R1 and R2 are in ohms, and C1 is in farads.

TONE MULTIPLEXING

Chapter 13 covered time domain multiplexing and which signals were sampled and then transmitted over a common media by interleaving the samples. If the timing is carefully controlled and synchronized at both ends, the total sampled data signals are recovered at the receiving end of the system. But time domain multiplexing, or TDM, is not the only form of multiplexing available. It is, however, preferred in *local* systems. Once we go outside of the immediate system and transmit data over radio or telephone facilities, some form of frequency domain multiplexing (FDM) is usually specified. Of course, high-density data communications systems may well use a combination of *both* TDM and FDM

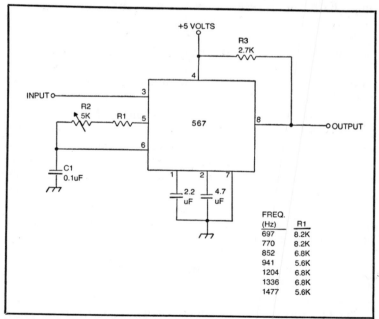

Fig. 14-17. PLL tone decoder.

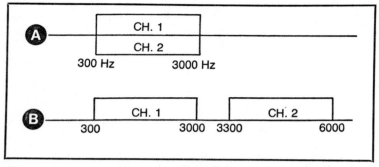

Fig. 14-18. Two channels of data occupy the same baseband at A, and frequency domain multiplexing of the second channel at B.

to greatly increase the bit rate (at, incidentally, the cost of a much greater required bandwidth).

A simplified version of FDM is shown in Fig. 14-18. In Fig. 14-18A, both channels 1 and 2 occupy the same baseband. They are both ordinary 300 - 3000 hertz audio communications channels. If these were transmitted over the same set of telephone lines, they would interfere with each other immensely. But if we can translate them to different frequencies, both can be sent over the same set of lines. Filters at the receiver end will recover them and split them off. In Fig. 14-18B, left channel No. 1 is in the baseband (300 to 3000 Hz) but shifted by heterodyning channel No. 2 into the range 3300 to 6000 Hz. The bandwidth occupied by channel No. 2 is the same 2700-Hz as it had when it was in baseband. But the upper and lower limits have been changed from 300 to 3000 Hz to 3300 to 6000 Hz.

In actual practice, audio grade lines can handle three—maybe four—shifted channels. Video grade lines, of course, can handle much more. But these figures are for voice communications channels or wideband data channels. Ordinary slow-speed to medium-speed data channels can use only two frequencies for each data bit (HIGH/LOW = F2/F1). While the actual bandwidth required will be determined by the speed of transmission, many more AFSK channels will fit into the same spectrum space occupied by only two or three voice channels.

Figure 14-19 shows a typical three-line data communications system based on FDM principles. Each channel uses a tone pair in which F1 and F2 are approximately 200-Hz apart. Channel No. 1 uses 1070 Hz for HIGH and 1270 Hz for LOW. Channel No. 2 uses 2025 Hz for HIGH and 2225 Hz for LOW, while channel No. 3 uses 2775 Hz for HIGH and 2975 Hz for LOW. Notice that we have inserted three, one-bit, data channels in the single-channel baseband space needed by voice communications. The tone outputs of all three encoders are applied to a combiner, usually a resistive network, and transmitted along the telephone or radio communications channel. At the receive end of the path, the signal composite is applied to the inputs of some high-Q filters. In some cases, a pre-filter will exist for each channel that will pass only the two desired frequencies. From then on, one of the decoder circuits shown previously will output a HIGH or LOW, depending upon which is required.

TELEPHONE LINES

The commercial telephone lines can now be accessed by non-ATT users. While it is necessary to subscribe and have the local telephone company install the required grade of line, it is no longer necessary for you to buy telephone company's equipment. For ordinary voice-grade lines, the bandwidth will be 300 to 3000 Hz. The transmitting source is expected to output approximately −5 VU of signal (VU is decibel scale referenced such that 0 VU is 1 mW in a 600-ohm load). The receiver is expected to tolerate signals in the −45 VU to −10 VU range.

Telephone lines tend to be long, so their ohmic resistance is significant. Fig. 14-20A shows an equivalent circuit, neglecting capacitance and inductance effects. The receiver looks like a resistance on the end of the line, while the R_{line} is the cummulative series resistance of the telephone line. When the receiver is on the

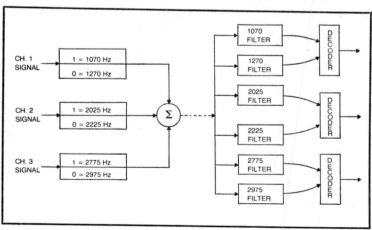

Fig. 14-19. Multi-channel frequency domain multiplexer.

Table 14-3. UART Pinouts.

Pin No.	Mnemonic	Function
1	Vcc	+5 volts DC power supply.
2	Vee	-12 volts DC power supply.
3	GND	Ground.
4	RRD	Receiver Register Disconnect. A high on this pin disconnects (i.e., places at high impedance) the receiver data output pins (5 through 12). A low on this pin connects the receiver data output lines to output pins 5 through 12.
5	RB8	LSB
6	RB7	
7	RB6	
8	RB5	Receiver data output lines
9	RB4	
10	RB3	
11	RB2	
12	RB1	MSB
13	PE	Parity error. A high on this pin indicates that the parity of the received data does not match the parity programmed at pin 39.
14	FE	Framing Error. A high on this line indicates that no valid stop bits were received
15	OE	Overrun Error. A high on this pin indicates that an overrun condition has occurred, which is defined as not having the DR flag (pin 19) reset before the next character is received by the internal receiver holding register.
16	SFD	Status Flag Disconnect. A high on this pin will disconnect (i.e., set to high impedance) the PE, FE, OE, DR, and THRE status flags. This feature allows the status flags from several UARTs to be bus-connected together.
17	RRC	16 × Receiver Clock. A clock signal is applied to this pin, and should have a frequency that is 16 times the desired baud rate (i.e., for 110 baud standard it is 16 × 110 baud, or 1760 hertz).
18	DRR	Data Receive Reset. Bringing this line low resets the data received (DR, pin 19) flag.
19	DR	Data Received. A high on this pin indicates that the entire character is received, and is in the receiver holding register.
20	RI	Receiver Serial Input. All serial input data bits are applied to this pin. Pin 20 must be forced high when no data is being received.
21	MR	Master Reset. A short pulse (i.e., a strobe pulse) applied to this pin will reset (i.e., force low) both receiver and transmitter registers, as well as the FE, OE, PE, and DRR flags. It also sets the TRO, THRE, and TRE flags (i.e, makes them high).
22	THRE	Transmitter Holding Register Empty. A high on this pin means that the data in the transmitter input buffer has been transferred to the transmitter register, and allows a new character to be loaded.
23	THRL	Transmitter Holding Register Load. A low applied to this pin enters the word applied to TB1 through TB8 (pins 26 through 33, respectively) into the transmitter holding register (THR). A positive-going level applied to this pin transfers the contents of the THR into the transmit register (TR), unless the TR is currently sending the previous word. When the transmission is finished the THR→TR transfer will take place automatically even if the pin 25 level transition is completed.
24	TRE	Transmit Register Empty. Remains high unless a transmission is taking place, in which case the TRE pin drops low.
25	TRO	Transmitter (Serial) Output. All data and control bits in the transmit register are output on this line. The TRO terminal stays high when no transmission is taking place, so the beginning of a transmission is always indicated by the first negative-going transition of the TRO terminal.
26	TB8	LSB
27	TB7	
28	TB6	
29	TB5	Transmitter input word.
30	TB4	
31	TB3	
32	TB2	
33	TB1	MSB

Continued on next page.

Table 14-3. UAR Pinouts (continued from page 210).

34	CRL	Control Register Load. Can be either wired permanently high, or be strobed with a positive-going pulse. It loads the programmed instructions (i.e., WLS1, WLS2, EPE, PI, and SBS) into the internal control register. Hard wiring of this terminal is preferred if these parameter never change, while switch or program control is preferred if the parameters do occassionally change.
35	PI	Parity inhibit. A high on this pin disables parity generation/verification functions, and forces PE (pin 13) to a low logic condition.
36	SBS	Stop Bit(s) Select. Programs the number of stop bits that are added to the data word output. A high on SBS causes the UART to send two stop bits if the word length format is 6, 7, or 8 bits, and 1.5 stop bits if the 5-bit teletypewriter format is selected (on pins 37-38). A low on SBS causes the UART to generate only one stop bit.
37 38	WLS1 WLS2 }	Word Length Select. Selects character length, exclusive of parity bits, according to the rules given in the chart below: Word Length — WLS1 — WLS2 5 bits — low — low 6 bits — high — low 7 bits — low — high 8 bits — high — high
39	EPE	Even Parity Enable. A high applied to this line selects even parity, while a low applied to this line selects odd parity.
40	TRC	16 × Transmit Clock. Apply a clock signal with a frequency that is equal to 16 times the desired baud rate. If the transmitter and receiver sections operate at the same speed (usually the case), then strap together TRC and RRC terminals so that the same clock serves both sections.

hook, its internal switch is open, so the voltage (see Fig. 14-20B) is approximately 60V. But when the receiver is lifted off the hook, the switch is closed and R_{rcvr} is in the circuit. This will load the circuit considerably. The voltage across the telephone receiver is now:

$$E = \frac{(60V) (R_{rcvr})}{(R_{line} + R_{rcvr})}$$

In most cases, it will be in the 6V to 26V range.

Figure 14-21 shows a method used in radio stations, police communications, and some amateur radio stations to connect the telephone line to receivers and transmitters. The same principle is also used in data communications using AFSK techniques. This circuit is called a phone patch. It is essentially a bridge that mixes the audio from the transmitter and the audio to the receiver in a linear but compensated network. The telephone line is one leg of the bridge. Note the use of the voice coupler. This is a protective device supplied by the telephone company. In the past, it was always necessary to rent a voice coupler from "Ma Bell." But it has recently become possible to buy one from a commercial outfit, provided that they guarantee it to meet telephone company specifications. Also, some local telephone companies have opened telephone stores that sell such accessories.

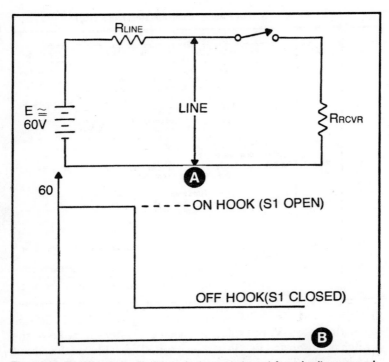

Fig. 14-20. Equivalent circuit of the telephone system at A, and voltage on and off hook at B.

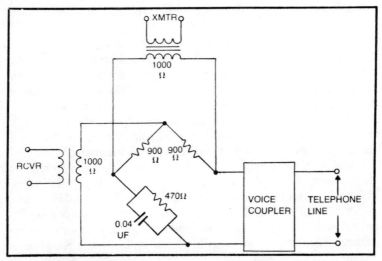

Fig. 14-21. Telephone patch.

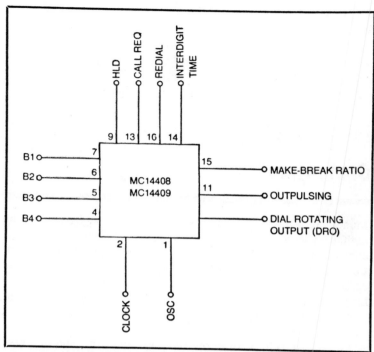

Fig. 14-22. MC14408 and MC14409 telephone dial pulser.

One last device to be shown, is the telephone dialer IC of Fig. 14-22. This is a special Motorola chip (MC14408 or 14409) that interfaces digital contact closure inputs will dial telephone circuits. Both this IC and the Touch-tone MC14410 can be directly controlled by a digital circuit or through a manual keyboard. Most of the automatic dialers will use such a circuit—usually the MC14410— and then output the desired numbers stored in binary registers.

Chapter 15
Computers

Immediately conjured up by the term "digital electronics" is the digital computer. In fact, to many people, digital electronics is computers and only computers (a too narrow view, as it turns out). Computers were introduced shortly after World War II. Those early monsters were vacuum tube affairs that consumed extremely large amounts of energy and were a constant maintenance technician's nightmare. Modern computers have shrunk in size while grown in capability because of microelectronics. The transistor was a giant leap forward in the computer industry, and the integrated circuit propelled computers even farther. Because of IC technology, an entire computer can be made on a single chip.

When we speak of computers, we usually refer to programmable digital computers. In fact, used by itself, the word "computer" is taken to mean only programmable digital computers. But there have been other types. Before World War II (although WW II accelerated their use due to gunnery needs), analog computers were known. These might be programmable or nonprogrammable (which actually means that someone else programmed it at the design stage, and then delivered the final, programmed product to you). An analog computer used analog operational amplifiers, dividers, multipliers, integrators differentiators, etc. to solve complex mathematical problems for which there is no easy algebraic method for obtaining a numerical answer (some differential calculus equations are said to be solvable only on an analog computer). Naval gunnery experts used both mechanical and electronic analog computers to aim a ship's guns at a target. The computer could work the problem much faster than a human and could automatically account for factors such as heat from firing distorting the long barrels of the guns.

Going even further back into history, we find that tabulating machines, often very analogous to "mechanical digital computers" were used even in the nineteenth century. The 1890 census, for example, was tabulated on such a machine. In fact, the IBM card used extensively in data processing today was actually a "Hollerith card," after the inventor of the 1890 census system. In the early 19th century, Babbage in England proposed a machine that would use punched cards to control weaving machines in the textile industry. Only the inability of technicians of the day to build the machine to the required tolerances prevented its operation. Stories circulate today, however, that someone built one of Babbage's machines on a model scale and it worked.

Today it is possible to buy a small desk top computer for less than the price of a new car (a lot less in many instances!) that will out perform machines current only 10 years ago. In that light, we will describe computers in terms of these microcomputers. Note please, that the ideas presented also apply to large mainframe and minicomputers. Only the specific technology is terribly different.

MICROCOMPUTERS AND MICROPROCESSORS

The microprocessor is a digital computer in the form of a large-scale integration (LSI) integrated circuit. Most of the currently popular microprocessor chips require a few additional support chips to make an actual digital programmable computer. The IC device called a *microprocessor* in those cases is actually just one part of the digital computer; i.e. the central processor unit (CPU). But newer devices, such as the Zilog, Inc. Z8000/Z8 types, actually contain some on board memory and I/O ports, so the device will operate as a complete programmable digital computer.

To the technician in the digital electronics industry, the introduction of microprocessor chips has obvious implications. They will only strengthen their service/installation market, already good and getting better. But how does it affect technicians in other fields? Plenty. It seems that a computer is very nearly a "universal machine." Clever programming, coupled with appropriate external circuitry and interface circuits, will allow the microprocessor to perform almost any job.

The usual view of most people regarding computers is that they are large megaliths, housed in top secret air conditioned rooms, that are used to make mistakes on your monthly utility bills or bank statements. But the truth is that such machines are only one type of computer. Another computer, still a programmable

Fig. 15-1. Small microcomputer system.

digital computer with a strong family resemblance to the large beast, may be housed on a 4×5-inch printed circuit board stuck into a corner of a microwave oven, or inside your new heat pump. In still another case, it might be a small desk top system (Fig. 15-1) used to control inventory, make payroll, and keep books for a small business firm.

We are seeing literally hundreds, possibly thousands, of new devices based on microcomputer/microprocessor technology. In my own field, medical instruments, the invasion of the microprocessor has become so pronounced that I suspect analog circuitry designers who have failed to upgrade will soon feel the employment pinch. It is now recommended that electrical engineering students take at least one microprocessor design course in order to make themselves more attractive to employers upon graduation.

With so many computers on the market, there has developed a massive service problem. Some manufacturers have become particularly hard hit in this area. It is unlikely that a customer whose business depends upon a microprocessor-based product will be inclined to generosity when it is found that the device must be air freighted back to the factory, or to some too-distant regional service center for repair. Nor is it advantageous for the manufacturer to have to admit that the nearest "local" service is 200 miles away, and the technician covers six states! At $35 per hour "portal-to-portal," such service is extremely costly. The editor of *Electronic Technician/Dealer* (ET/D) magazine, a forward-thinking leader in electronic trade publications, queried a number of smaller digital equipment makers about their intentions for servicing their products. The reply was that local, independent businesses would be contracted to service the equipment. They were including TV/audio repair shops in this intention. With modern PC board replacement, only a small amount of extra training will be needed for such technicians. The idea is to have local businesses make the initial call. If the device cannot be repaired at the component level within a certain time period, then the device or the affected PC board is to be replaced; the original going back to the shop or factory for repair. With the incidence of TV service jobs dropping nationally, this is one good alternative for the small electronic service business struggling to stay in business. One authority claims that the repair incidence dropped from 2.4 repairs per set in 1969 to 0.6 repairs per set in 1977.

WHAT IS A MICROPROCESSOR?

A *microprocessor* is an LSI integrated circuit (usually in a 40-, or more, pin DIP IC package) that contains most or all of the circuitry needed to make a programmable digital computer. Many people erroneously use the terms microcomputer and microprocessor as synonyms. They are not synonyms. A microcomputer is a digital programmable computer using one or more microprocessor chips. The microprocessor is just an IC, while the microcomputer is a complete computer.

Nor is the microprocessor a calculator chip. Many people seem to think that the microprocessor and microcomputer are merely calculators with a lot of junk hanging on. But these technologies are different. While they are cousins—even kissing cousins—they are very different. The similarity stems partly from the digital technology used to implement the two devices, and partly from the fact that both can calculate. Some of the more modern calculators are very powerful and look much like a computer. If this trend continues, the distinction between them may well narrow. But there is a principle difference. A calculator is a special-purpose device, even the programmable models, while a true computer is a universal or general-purpose machine. The calculator receives its very limited instructions only through keyboard contact closures, each of which has a special unique significance. The difference that seems to be changing is in the matter of programmable calculators.

WHAT CAN A COMPUTER DO?

When a computer hobbyist or technician seeking to upgrade skills buys a microcomputer, the very first question that will often be asked is, "But, *what will it do?*" Except for the fact that the questioner doesn't know the answer, this question would be simply absurd because it is too broad for even an attempt at an intelligent answer. Almost any job that can be performed by a sequence of small steps can be performed by a cleverly programmed computer. The range of jobs *now* being done by these devices is almost endless. The range runs from simple light controls to factory process controllers to complex data processing numbers crunching applications normally identified as a computer job.

Some computers in no way resemble the popular notion of how a computer should look. For example, remember all of those video games that tempted you last Christmas season? Almost all of them now use microprocessor technology instead of the older fixed digital circuitry. By this move, the games have acquired lower production cost, more flexibility, and a lot of simplicity in design. Changing games is a simple matter of changing instructions stored in permanent read only memory (ROM) . . . which you buy as "cartridges" like phono records or prerecorded tapes.

In almost all cases where a microprocessor or microcomputer has replaced discrete digital logic circuits, the simplicity of circuitry acquired is due to asking the program of the computer to replace some of the hardware previously used. The universality of the these devices is that new jobs or new games can be performed only by changing the ROM and insuring that the correct interface is available. The electronic games market serves our purpose here to show the universality of computers. The same principle that allows the user to select programs from a "library" or games also allows us to make the computer do a lot of different jobs. One of the few differences between the computer that controls a factory process (lathes, machines, etc.) and the computer in the factory's bookkeeping department is one of programming and interface. If all of the process control jobs can be performed through ordinary computer input/output ports, then the difference is merely the program stored in memory. It is not unlikely that a single computer would be used to perform process control during the day and then be pressed into service at night for administrative computing that could wait until the next day.

In entertainment electronics we find microprocessor designs all over the place. Several hi-fi and at least two auto radio manufacturers now use microprocessor circuitry to control the FM local oscillator in FM tuners. At least one company sells a stereo cassette deck that uses a microprocessor to do program control jobs.

Fig. 15-2. Single-board training computer.

Since the late 70s, the business applications of microprocessors has increased dramatically. This revolution has brought computer power to businesses too small only a few years ago to even contemplate computer ownership.

TYPES OF MICROCOMPUTERS

Now that we have established the fact that a microprocessor is a chip and a microcomputer is a computer and that they are general-purpose analytical engines, let us proceed to break the computers down even further. There are three basic forms of computer: process controllers (PC), single board computers (SBC), and mainframe systems (MS).

The PC and SBC devices are very similar to each other. Both are usually built on a single, small, printed circuit board. Two examples of single board computers are shown in Figs. 15-2 and 15-3. These machines, while capable of being expanded into very complex systems through the addition of external circuitry, are used primarily for training applications. They form a good means for low cost instruction and allow the student to learn the basics of microprocessors using a real computer. Registers and memory locations can be examined before and after simple programs are executed, allowing the student to see the effect of the various program instructions.

Most SBCs have keyboard entry, and output is an LED or liquid crystal display. Some also have a standard 20-mA teletypewriter loop for hard copy output. Many also have a built-in audio

Fig. 15-3. Single-board computer in a cabinet.

cassette interface that will allow the user to store and read programs for later use.

The single-board computers shown here are both housed in some type of package. The Imsai is built in a low cost plastic case that serves mostly to protect the sensitive electronic devices (CMOS is used extensively in microprocessors). The Heath uses a somewhat more attractive case and is a desk top model. Many SBCs, however, are sold bare with no case—only a PC board with LEDs and keyboard.

The process controller is much like an SBC in that it is usually housed in a single printed circuit board. It is, however, not usually equipped with LED readouts and keyboard. These computers are designed to be integrated into other electronic equipment. Examples are the National Semiconductor SC/MP and the Pro-Log machines. Zilog and Mostek (Z-80 manufacturers) also provide PC versions of their product.

An example of another class of SBC is shown in Fig. 15-4. This is the Digital Equipment Corporation (DEC) LSI-11. It responds to the commands also honored by DEC's very powerful PDP-11-series minicomputers. Heathkit uses a drop-in LSI-LL DEC card inside of their H11 microcomputer, thereby gaining PDP-11 software compatibility. Other companies are also offering this class of SBC. Note that they are very much like mainframe minicomputers, without the mainframe. In applications requiring a very powerful computer as a PC, then the designer might opt for something like the LSI-11 as a lower cost alternative to designing a new computer.

SBC/PC Memory Limitations

Most typical PC and SBC microcomputers have a very limited memory, so can support limited applications. They may have from 1 to 2 kilobytes of random access read/write memory and from 256 bytes to 8K bytes of read only memory (ROM). There is a significant difference between RAM and ROM memory. The RAM allows the CPU to read data from, or write data to, memory locations. But the ROM is preprogrammed and allows the CPU to only read data from memory. In any given commercial applications, there may be RAM and ROM mixed in the same device.

One result of memory limitations in SBC/PC devices is that we must program in *machine language*; the binary words recognized as instructions by the computer. Any given microprocessor or microcomputer will have an instruction set of binary numbers that the computer will recognize as valid instructions to perform certain specific tasks. Machine language instructions, however, are tedious to enter, even in relatively simple programs. Most businesses require a *high level language* (BASIC, FORTRAN, COBOL) that is more like either English or algebra. Such programs can be entered on a keyboard or video terminal, and are a lot easier for non-computerniks to understand. It simplifies the chore of business programming at the expense of memory and processing

Fig. 15-4. DEC LSI/11 microcomputer.

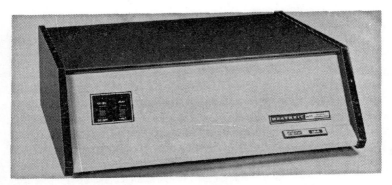

Fig. 15-5. Heath H11 computer uses the DEC LSI/11 module.

time efficiency. The machine language is more difficult to program in, but is much easier on the memory space required and the time of execution.

Mainframe Microcomputers

The term *mainframe computers* is usually applied to the large IBM 370-class of machine, but is equally applicable to smaller desktop units. Figures 15-5 and 15-6 show a simple desk top mainframe computer. The machine in Fig. 15-6 contains its own video display (identical to a TV receivers from the video detector to the output, less only the rf/i-f sections). Programming and data entry are performed through a keyboard. In some models, the keyboard is attached, while in others it is semiportable.

Other accessories used with mainframe computers include teletypewriter or printer for hardcopy printout, cassette tape (audio and digital) or floppy discs for mass storage of data and programs, and a variety of other peripherals.

Most mainframe computers of any size will use a mother board with plug in printed circuit cards. The motherboard contains the power supply distribution, and the data/address buses that communicates between cards. One of the most popular, although by no means the most universal, mainframe buses is the so-called S-100 bus, originally designed for the Altair machine. In the past, it was easy to categorize the boards used in bus-oriented computers. But today, it is not too easy to make such statements. There are a number of single board designs in which a CPU, video, cassette, and I/O functions, plus some limited memory, is build onto a single S-100 board.

There are other approaches to mainframe design, as well. The APPLE computer, based on the 6502 microprocessor chip uses a

222

single board for CPU, 40K bytes of memory, and some other functions. But it is also a motherboard, because it accepts several plug-ins that are either memory of I/O based. A typical set-up might consist of one slot filled with a ROM board programmed in APPLE BASIC, with several other slots filled with I/O prots or device controllers.

HOW DOES A COMPUTER WORK?

The very first thing that I am going to do in this section is to cop out—the subject would take a collection of several books to do justice. In that vein, let me first point out at least one book that I can recommend from personal experience: TAB book No. 785, *Mircorproccessor/Microprogramming Handbook* by Brice Ward.

A simplified block diagram to a computer is shown in Fig. 15-7. The *arithmetic logic unit* (ALU) performs the arithmetic and logical functions (AND, OR, NAND, NOR, XOR, control), while much of the intrachip and interchip control function is handled by the *control logic* section. The control section is used to issue status flags that tell the outside world what is happening right now, what just happened, or what the result of an arithmetic operation was (negative/positive number, etc). There are also several general-purpose and special-purpose registers inside of the CPU.

The main register, the one referred to in most of the instructions, is the accumulator (also called the A register in some machines). All data entering or leaving the CPU via the eight-bit data bus must pass through the accumulator first.

The *instruction register* will contain the last instruction fetched from memory, while the *instruction decode* section examines the

Fig. 15-6. Mainframe microcomputer.

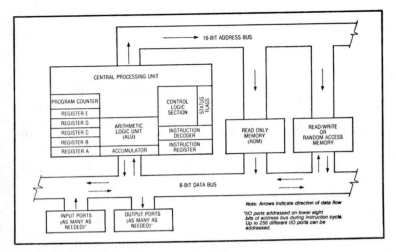

Fig. 15-7. Block diagram of a typical computer.

contents of the instruction register and determines what the computer is supposed to do next. The program counter (PC) is a special-purpose register than controls the fetching of instructions. It is initially loaded with the *address* of the first instruction to be executed. This instruction will be stored somewhere in memory, and the PC will contain the address of the memory location at which it is found. After each instruction is executed the program counter will either increment to pick up the instruction at the next sequential memory location or be loaded with an alternate next instruction address. The latter course is sometimes dictated by the particular instruction just executed; i.e., it could call for a branch, or would occur in response to an *interrupt* request from the outside world.

The *reset* function on a computer is usually a switch that, when closed, cause an immediate jump to some particular location in memory (usually ØØØØØØØØ ØØØØØØØØ). The instruction at that location may well be the first instruction in the program, in which case the program counter would be located with ØØ ØØ (hex). But it could also be a jump instruction that tells the PC to load the address given in the instruction at ØØ ØØ hex. The next instruction, then, would be found at the location now loaded into the PC.

Memory

One of the significant advances in computing machinery was the invention of the idea of *memory*. The memory is a section of locations (flip-flops that can be ordered to set or reset) that will contain either *data* or *instruction operation codes*. An analogy to the computer memory is the pidgeon holes used by your local mail carrier to sort mail. The letters represent data, while the pigeon holes represent individual houses—locations with specific addresses. The sorter (CPU) will store the data at the locations called for by the address.

In a computer, the pigeon holes are actually ordered arrays of ordinary flip-flops. The condition of the flip-flop (1 or $\emptyset$) is determined according to the type of data to be stored. To make an eight-bit (one-byte) memory location, we need eight flip-flops; each FF represents one bit.

Machine Cycles

Computer instructions are binary words. Coded numerical commands called op-codes tell the control section of the CPU what is to be done. But data is also in the form of a binary number. In the Zilog, Inc Z-80, for example, the op-code for an input-immediate instruction is DB (hec), or 11011011. But 11011011 is also the binary number representation for 219_{10}. The computer cannot tell which is intended unless we give it some help. After, all, both instruction 11011011 and data 11011011 will be input through the accumulator and data bus. To overcome this problem, programmable digital computers operate in *machine cycles*. There will be at least two such cycles: *instruction fetch* and *instruction fetch*. During

Fig. 15-8. PET microcomputer.

Fig. 15-9. Sol Terminal microcomputer.

Fig. 15-10. Another popular microcomputer.

226

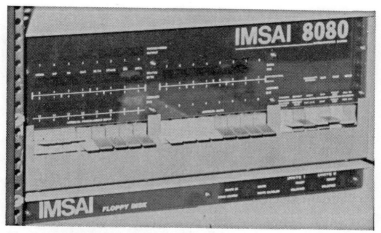

Fig. 15-11. IMSAI 8080 microcomputer.

Fig. 15-12. IMSAI model with the covers removed to show the mother board and power supply.

Fig. 15-13. IMSAI mainframe computer with dual disc drive and TV readout.

the fetch cycle, the computer will go to the memory location specified by the program counter, and enter the binary word found there into the accumulator. From there it is transferred to the instruction register, and then decoded. The decoding and carrying out of the instruction is done during the execution cycle. Following execution, the program counter will be incremented by one address location, or will be loaded with a new word that was determined by the instruction just executed. The new word in the PC tells the computer where the next instruction is to be found. The computer will alternate between instruction fetch and execution cycles continually.

In the next chapter we will examine a particular microprocessor CPU chip, the Zilog, Inc. Z-80. The reason this was selected is that it is popular, and the author knows a lot about it . . . my own hobby microcomputer is Z80-based. But first, examine Figs. 15-8 through 15-13.

Chapter 16
The Z80—A Typical Microprocessor

The age of the microprocessor chip began in the 1971 when Intel introduced the 8008 device. The 8008 is crude by today's standards and was almost immediately supplanted by another Intel offering, the 8080. The newer device was to become one of the most popular microprocessor chips on the market for a long time. The Z80, which is manufactured by Zilog, Inc., with Mostek as a second source, uses all of the 8080 instructions and adds a few of its own. But the Z80 is not merely an updated version of somebody else's toy; it is a completely new device in its own right. Do not think that the Z80 is the be-all and end-all of all microprocessors, however. The reason that this device was selected as our example in this elementary text is that the author is familiar with the Z80, more so than 8080, 6502 or 6800 devices.

The internal structure of the Z80 is not much unlike the block diagram of the last chapter. Indeed, many computers could use similar block diagrams. The Z80 device uses an eight-bit data bus and a sixteen-bit address bus. This configuration means that the Z80 will address up to 2^{16}, or 65536, different memory locations. The Z80 has 18 special-purpose or general-purpose eight-bit registers, and four 16-bit special-purpose registers. The main register set is grouped into two groups: the main register set and the alternate register set.

SPECIAL-PURPOSE REGISTERS

The special-purpose registers are the program counter (PC), the stack pointer (SP), index registers IY and IX, interrupt vector (I), and memory refresh (R).

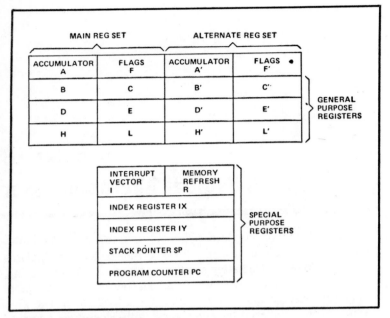

Fig. 16-1. Z80 register arrangement.

Program Counter

As in all computers, the PC holds the 16-bit address of the instruction currently being fetched from a memory location. The program counter contents are normally either incremented by one count or is loaded with the memory address called for by a jump instruction.

Stack Pointer

A stack is a section of external memory organized as a *last-in-first-out* (LIFO) file. Data can be pushed onto the stack or popped off of the stack. The SP register contains the 16-bit address of the *top* location of the external memory stack.

Index Registers

Indexed addressing uses the contents of an index register and a byte stored in memory one location away from the op-code for an index addressing instruction in order to calculate the address of a memory location. The Z80 contains two index registers: IX and IY.

Interrupt Address Register (I)

The Z80 uses obe address mode in which the effective location of an instruction is given by the combination of two bytes that form a two-byte, or 16-bit, word). One byte (high order) is supplied by the contents of the I register, while the other is supplied by the interrupting device.

Memory Refresh Register (R)

A certain type of external memory device, called dynamic memory, requires periodic refreshing. The R register contains data to refresh the memory.

GENERAL-PURPOSE REGISTERS

The Z80 contains two accumulators, two flag registers, and six general-purpose registers. The accumulator, of course, is the

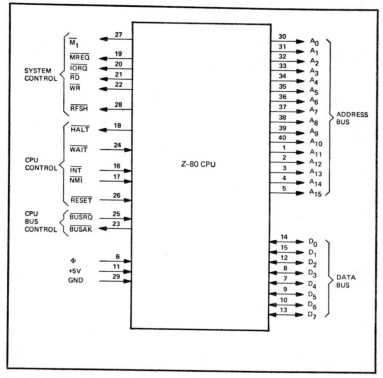

Fig. 16-2. Z80 pinouts and definitions.

Table 16-1. Z80 Pin Functions.

A₀-A₁₅ (Address Bus)	Tri-state output, active high. A_0 - A_{15} constitute a 16-bit address bus. The address bus provides the address for memory (up to 64K bytes) data exchanges and for I/O device data exchanges. I/O addressing uses the 8 lower address bits to allow the user to directly select up to 256 input or 256 output ports. A_0 is the least significant address bit. During refresh time, the lower 7 bits contain a valid refresh address.
D₀-D₇ (Data Bus)	Tri-state input/output, active high. D_0-D_7 constitute an 8-bit directional data bus. The data bus is used for data exchanges with memory and I/O devices.
$\overline{M_1}$ (Machine Cycle one)	Output, active low. $\overline{M_1}$ indicates that the current machine cycle is the OP code fetch cycle of an instruction execution. Note that during execution of 2-byte op-codes, $\overline{M_1}$ is generated as each op code byte is fetched. These two byte op-codes always begin with CBH, DDH, EDH or FDH. $\overline{M_1}$ also occurs with $\overline{IORQ}$ to indicate an interrupt acknowledge cycle.
$\overline{MREQ}$ (Memory Request)	Tri-state output, active low. The memory request signal indicates that the address bus holds a valid address for a memory read or memory write operation.
$\overline{IORQ}$ (Input/Output Request)	Tri-state output, active low. The $\overline{IORQ}$ signal indicates that the lower half of the address bus holds a valid I/O address for a I/O read or write operation. An $\overline{IORQ}$ signal is also generated with an $\overline{M_1}$ signal when an interrupt is being acknowledged to indicate that an interrupt response vector can be placed on the data bus. Interrupt Acknowledge operations occur during $\overline{M_1}$ time while I/O operations never occur during $\overline{M_1}$ time.
$\overline{RD}$ (Memory Read)	Tri-state output, active low. $\overline{RD}$ indicates that the CPU wants to read data from memory or an I/O device. The addressed I/O device or memory should use this signal to gate data onto the CPU data bus.
$\overline{WR}$ (Memory Write)	Tri-state output, active low. $\overline{WR}$ indicates that the CPU data bus holds valid data to be stored in the addressed memory or I/O device.
$\overline{RFSH}$ (Refresh)	Output, active low. $\overline{RFSH}$ indicates that the lower 7 bits of the address bus contain a refresh address for dynamic memories and the current $\overline{MREQ}$ signal should be used to do a refresh read to all dynamic memories.
Halt (Halt state)	Output, active low. HALT indicates that the CPU has executed a $\overline{HALF}$ software instruction and is awaiting either a non maskable or a maskable interrupt (with the mask enabled) before operation can resume. While halted, the CPU executes NOP's to maintain memory refresh activity.
$\overline{WAIT}$ (Wait)	Input, active low. $\overline{WAIT}$ indicates to the Z-80 CPU that the addressed memory or I/O devices are not ready for a data transfer. The CPU continues to enter wait states for as long as this signal is active. This signal allows memory or I/O devices of any speed to be synchronized to the CPU.

Table 16-1. Z80 Pin Functions (continued from page 232).

INT (Interrupt Request)	Input. active low. The Interrupt Request signal is generated by I O devices. A request will be honored at the end of the current instruction if the internal software controlled interrupt enable flip-flop (IFF) is enabled and if the BUSRQ signal is not active. When the CPU accepts the interrupt. an acknowledge signal (IORQ during M1 time) is sent out at the beginning of the next instruction cycle. The CPU can respond to an interrupt in three different modes that are described in detail in section 5.4 (CPU Control Instructions).
NMI (Non Maskable Interrupt	Input. negative edge triggered. The non maskable interrupt request line has a higher priority than INT and is always recognized at the end of the current instruction. independent of the status of the interrupt enable flip-flop. NMI automatically forces the Z-80 CPU to restart to location 0066H. The program counter is automatically saved in the external stack so that the user can return to the program that was interrupted. Note that continuous WAIT cycles can prevent the current instruction from ending, and that a BUSRQ will override a NMI.
RESET	Input, active low. RESET forces the program counter to zero and initializes the CPU. The CPU initialization includes: 1(Disable the interrupt enable flip-flop 2) Set Register I = 00H 3) Set Register R = 00H 4) Set Interrupt Mode 0 During reset time. the address bus and data bus go to a high impedance state and all control output signals go to the inactive state.
BUSRQ (Bus Request)	Input, active low. The bus request signal is used to request the CPU address bus, data bus and tri-state output control signals to go to a high impedance state so that other devices can control these buses. When BUSRQ is activated. the CPU will set these buses to a high impedance state as soon as the current CPU machine cycle is terminated.
BUSAK (Bus Acknowledge)	Output, active low. Bus acknowledge is used to indicate to the requesting device that the CPU address bus. data bus and tri-state control bus signals have been set to their high impedance state and the external device can now control these signals.
φ	Single phase TTL level clock which requires only a 330 ohm pull-up resistor to +5 volts to meet all clock requirements.

main register. The flag register is used to contain the bits that indicate the status of the CPU, or the result of the operations.

The general-purpose registers are labeled B,C,D,E,H, and L in the main register set, and B', C', D', E', H', and L'. The two accumulators are labeled A and A', while the two flag registers are F and F'. Figure 16-1 summarizes the Z80 register set.

Figure 16-2 shows the Z80 pinouts. The Z80 is packaged in a 40-pin DIP IC package. The pin functions are summarized in Table 16-1.

Chapter 17
Memory-I/O Interfacing in Computers

In most digital computers or other instruments busses are used to transfer data or provide addresses to locations. These busses are parallel data paths between sections of the computer. But because all sections share the same bus, some means must be provided for decoding specific input/output port or memory location addresses. There are, in most eight-bit microcomputers, 256 different allowable I/O ports, and 65,536 memory locations! Obviously, in order to avoid chaos, we must tell the memory:

□ That a memory operation is taking place.

□ Whether it is a *read* or *write* operation.

□ Which of 65,536 locations is intended in this operation. (The first two steps are combined in some microprocessors.)

Similarly, for I/O operations, we need do the same thing, except that the first step is modified to indicate an I/O operation:

□ Specify one of 256 possible (000 to 255) devices or I/O ports is intended.

□ Indicate to the I/O ports that an I/O operation is to take place.

□ Specify whether the operation is an input (read) or output (write).

A *device selector* is used to turn on, or signal, some device or circuit external to the CPU. Basically, most (perhaps all) such devices are seen by the CPU as either *memory* or *I/O ports*. The device select function, then, is essentially the same as these functions.

BINARY WORD DECODERS

Binary word decoder circuits are designed to provide an output indication when a specified binary word is applied to its input. These circuits are used to decode the data on the address bus to determine which of 65,536 memory locations is intended, or which of 256 I/O ports are intended.

In simple, four-bit applications we can use some of the decoder and data distributor ICs in the 7400-series (TTL) line. We may also use the 7485 four-bit comparator.

Figure 17-1 shows the use of 7442 1-of-10 BCD decoder. This device is normally used to select the vertical column decade light, or *Nixie*® tube elements that correspond to the four-bit (BCD) binary word at the inputs. Note that the normal range of four-bit binary is 0 to 15 decimal. This means that only part of the range is usable; i.e., 0 to 9. Within that constraint, however, we can find much application for the 7442 device. When the selected binary word is applied to the inputs (BØ - B3), then the corresponding output drops LOW. As an example, let us say that we want to designate a given device or circuit as input device No. 6 ($6_{10} = 0110_2$). Since 0110 is within the range of the 7442, we can

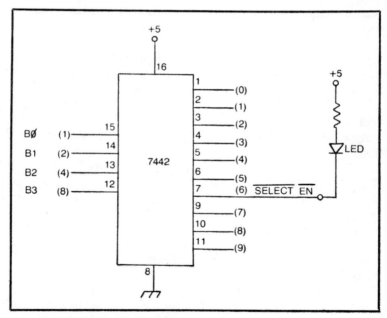

Fig. 17-1. Four-bit decoder using 7442 device.

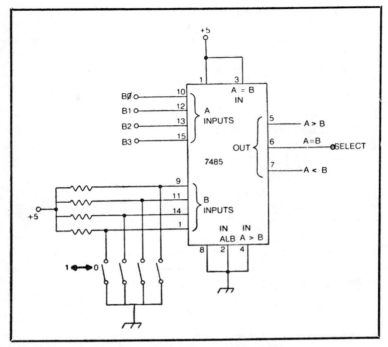

Fig. 17-2. Programmable decoder using the 7485 device.

use that IC instead of some more expensive device. In our simplified example, we will use an LED (light-emitting diode) to simulate a peripheral device that wants to see a LOW on the *enable* (EN) input to turn on. The LED remains turned off if any binary word other than 0110 is applied to BØ-B3. But if 0110 is seen, then output No. 6 (pin No. 7) of the 7442 goes LOW, turning on the LED.

The use of the 7485 data comparator is shown in Fig. 17-2. This TTL device examines two, four-bit binary words (designated A and B), and issues outputs indicating one of three possible states: A equals B, A less than B, A greater than B. In this application, we are interested in A=B. Note that there are also three inputs for the same three conditions. We tie A > B and A < B to ground and A=B to +5 VDC (logical-HIGH). We will program the B inputs with the number in binary (0000 to 1111, or 0 to 15) assigned to the device. If, as in the previous example, we call the device port No. 6, we will load 0110 into 7485 input B. This job is accomplished in the example circuit of Fig. 17-2 by switches, but would probably be

236

hand-wired in most equipment. When 0110 is applied to the A-inputs, the A=B condition exists, so pin No. 6 goes HIGH. In this circuit, the select pulse output is active-HIGH, so will turn on only those devices that want to see a positive-going enable pulse. If an active-LOW enable pulse is needed, then an inverter at the 7485 A=B output will be required.

A one-of-16 decoder is shown in Fig. 17-3. This circuit is based on the 74154 TTL data distributor IC. By changing the level applied to pin No. 18, we can determine whether the selected output is active-HIGH or active-LOW. A HIGH on pin No. 18 produces the active-HIGH mode, while a LOW on pin No. 18

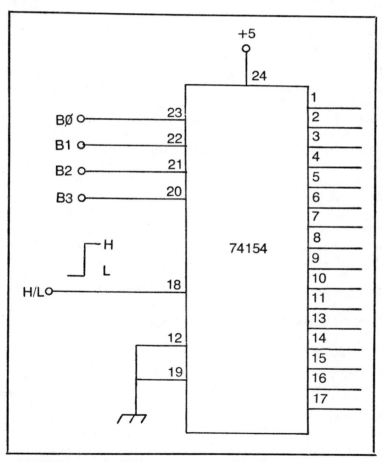

Fig. 17-3. Four-bit 1-of-16 decoder using 74154 device.

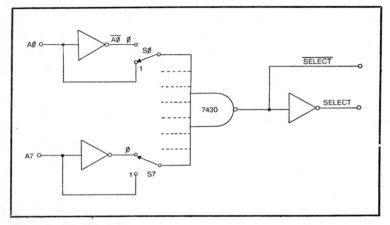

Fig. 17-4. Eight-bit decoder.

produces the active-LOW condition. The input is four-bit binary, allowing up to 16 counts. Corresponding to the 16 unique states in four-bit binary (0000 to 1111). The device enable line on the peripheral will be connected to the 74154 output corresponding to the 16 unique states in four-bit binary words (0000-1111). The 74154 also allows selection of active-HIGH/active-LOW output; not possible on the 7442. The 74154 is a 24-pin DIP device and requires somewhat more space on a printed circuit board than does the 7442.

Thus far, our discussion has centered around four-bit data words. But this is legitimate only when less than sixteen choices are needed; i.e., when all addresses can be given in the number range of the four-bit word (less than 16 choices). But most microcomputors, and those digital instruments that are based on the microprocessor chip, are oriented around eight or 16 bit data and address buses. If we wish to take full advantage of these devices, then we must be able to decode word lengths longer than four bits.

EIGHT-BIT WORD DECODERS

An eight-bit binary word decoder circuit used in many microcomputer designs is shown in Fig. 17-4. This circuit is based on the 7430 eight-input TTL NAND gate IC. Recall the operation of a NAND gate: if any one input is LOW, then the output is HIGH. It requires all inputs (in the case of the 7430, all eight) must be HIGH the output to be LOW. In this type of decoder, we want to connive to make all eight inputs of the NAND gate HIGH when the required

238

word appears on bits AØ through A7 of the address bus. But there is only one word (1111) that has all HIGH bits naturally. We must use inverters to flip the normally LOW bits of the desired word to HIGH. Only two inverters are shown here (at the LSB and MSB positions), but you may assume as many as eight, if needed.

Let's consider the AØ inverter. If the bit at that position in the desired binary word will be HIGH, then swtich SØ will be in the "1" position. The AØ line is now connected directly to one input of the NAND gate. But what happens when the anticipated AØ bit is LOW? Here we must set switch SØ to the "Ø" position. Line AØ is not connected directly to an input of the NAND gate, but is first inverted; AØ is applied to the AØ input of the NAND gate. If all switches are correctly set, then the nand gate will see the binary word 11111111 *only* when the correct word is applied to AØ-A7.

Figure 11-5 shows an actual example of a decoder. This particular circuit could be used to decoder I/O port addresses (in many microcomputers the eight-bit I/O port address appears on the lower byte of the address bus, hence the AØ through A7 designations). This circuit is set to decode the binary word 11010011. Bits AØ, A1, A4, A6, and A7 will be "1" (HIGH), so those lines are connected directly to the 7430 NAND gate. Bits A2, A3 and A5 will be "Ø" (LOW) so they must be inverted prior to being applied to the 7430 inputs. The only binary word an AØ - A7 that will produce 11111111 at the 7430 inputs, then, is 11010011, the required word.

The circuits of Figs. 17-4 and 17-5 have proven extremely popular with designers of digital equipment. They can boast both low cost and simplicity.

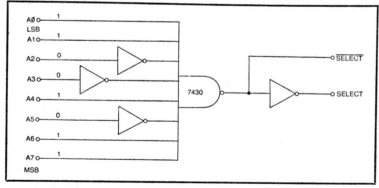

Fig. 17-5. Decoder for the word 11010011.

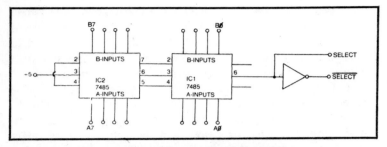

Fig. 17-6. Cascading 7485 devices to make eight-bit decoder.

An eight-bit circuit based on the 7485 TTL comparator IC is shown in Fig. 17-6. The 7485 is a four-bit device, so two must be used in cascade to accommodate eight-bit situations. This is done by connecting the outputs of one stage to the *cascade inputs* of the following chip. IC1 will not output a HIGH unless the A=B criterion is met by both chips.

The operation of this circuit is essentially the same as in the four-bit version. The desired address is loaded into one set of inputs, while the bus being monitored is connected to the other. Pin No. 6 goes HIGH *only* if A=B. At all other times pin No. 6 is LOW.

An extension of the 7442 circuit is shown in Fig. 17-7. Recall that the 7442 is a BCD-to-1-of-10 decoder. It will produce a LOW output that corresponds to any decimal equivalent of binary from 0000 to 1001 (0 to 9 in decimal). By using two 7442 devices and a single, two-input, 7402 NOR gate we can generate many more combinations (ØØ to 99). The inputs to the 7442 are connected in four-bit "nybbles" (half a byte) to the eight-bit bus. The inputs to the 7402 are connected to the output corresponding to the desired code. Assume that IC1 is the low order half-byte, and that IC2 is the high order half-byte. Let us further assume that we assign output port No. 26 to this circuit (see Fig. 17-8).

When the code on the address bus is 00100110, then the "2" output of IC2 and the "6" output of IC1 will both drop LOW simultaneously. Recall the rules governing a TTL NOR gate. The output is HIGH only if both inputs are LOW. This condition, then, generates a select pulse at the output of the NOR gate.

I/O SELECT CIRCUITS

Selecting an input/output port requires more than an address decoder. The address bus is also used for memory addressing, so

240

to avoid confusion, we must be able to recognize the I/O state of the CPU. In this section, we will describe typical circuits using the Z80 microprocessor IC control signals. Keep in mind that other microprocessor chips will use slightly different control signals, or will use different names for the same signals.

The Z80 uses two separate pins for read and write signals. The active-LOW outputs are designated $\overline{RD}$ and $\overline{WR}$, respectively. But the $\overline{RD}$ and $\overline{WR}$ terminals are also used for memory operations. so we also need another set of control signals to differentiate between these functions. In the Z80 device, there are separate *input/output request* (IORQ) and *memory request* (MREQ) signals to service that function. When an I/O operation is to take place, the IORQ goes LOW. But if a memory operation is taking place instead, then MREQ goes LOW.

Figure 17-9 shows a circuit that will generate OUT and IN directional commands when the desired I/O port address is presented on the lower eight bits of the address bus. The circuit is built around the 7402 device, a TTL NOR gate. The basic operating rules of the NOR gate are: both inputs of the NOR gate must be LOW in order to generate a HIGH output. The output will be LOW if either NOR gate input is HIGH. Gates G1 and G2 can only be enabled when IORQ is LOW (active) and the appropriate output (inverted) is received from G3 and G4, respectively.

Let us assume an *output* operation is to take place, and the correct port address has been received at the address decoder. The

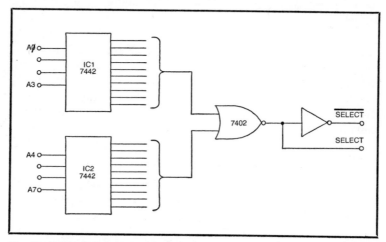

Fig. 17-7. Selectable eight-bit decoder.

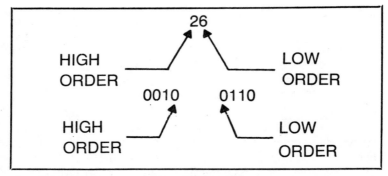

Fig. 17-8. Operation of the 7442 circuit of Fig. 17-7.

correct address will cause the $\overline{\text{SELECT}}$ line to drop LOW, while an output command causes $\overline{\text{WR}}$ (an output is a write operation) to drop LOW. Both inputs of G3 are now LOW, so its output goes HIGH. This condition forces the output of inverter NO. 1 LOW. Gate G1 now sees both inputs LOW because IORQ is also LOW at this time. The output of G1, then, snaps HIGH, generating the OUT signal.

Similar reasoning describes the creation of the IN signal; only now the operative gates are G2 and G4. Gate G2 is enabled when the IORQ lines goes LOW, indicating that an I/O operation is taking place. But the IN signal cannot be created until the $\overline{\text{RD}}$ line goes LOW (inputs are read operations). When both criteria are satisfied, then both inputs of G4 are LOW, and that makes the output of G4 HIGH. Now all three criteria for an input operation are satisfied, we can generate an IN signal. The output of gate G2 goes HIGH.

We may also use the 7442 device to generate an IN or OUT signal. A sample circuit is shown in Fig. 17-10. The 7442 BCD-to-1-of-10 decoder causes one of ten possible outputs to drop LOW in response to an appropriate four-bit binary word at the input. We can make I/O control signals of the Z80 look like a three-bit word. The fourth bit required by the 7442 is obtained by grounding the D input (pin No. 12) of the 7442. This makes the most significant bit of the code always LOW (logical-0). To generate a binary code, we connect the Z80 to the 7442 as follows:

7442 Input	Z80 Control Signal
A	IORQ
B	RD
C	WR
D	(Zero)

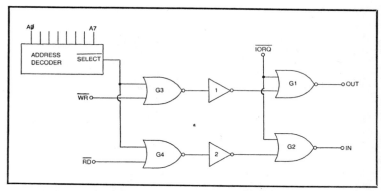

Fig. 17-9. Typical WR/RD circuit to generate IN and OUT commands.

The respective binary codes, then, will be:

Command	BCD	Decimal	7442 Pin No.
IN	0100	4	5
OUT	0010	2	3

We must also account for the I/O port address, as indicated by either SELECT or $\overline{\text{SELECT}}$ from the address decoder circuit. There are at least three common approaches to solving this problem. One of them is to use NOR gates, as shown in Fig. 17-10. The other two involve using the D-input of the 7442.

In Fig. 17-10, two NOR gates are used. One input will go LOW when $\overline{\text{SELECT}}$ from the address decoder goes LOW. The

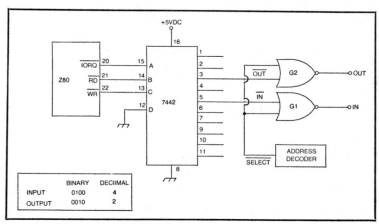

Fig. 17-10. Interfacing Z80 with 7442 decoder.

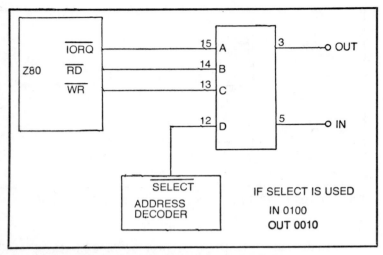

Fig. 17-11. Making better use of the D input of the 7442.

remaining pinouts are connected to the 7442 outputs that corres-
pond to IN and OUT. The operation is as follows:
Input operations:
   □ $\overline{IORQ}$ and $\overline{RD}$ go LOW, so the BCD word at the 7442
      input is 0100. This causes output No. 4 (pin 5) to go LOW.
      This pin 5 signal is the $\overline{IN}$ signal.
   □ At the same time, the port address appears on bits AØ
      through A7 of the address bus, and the address decoder
      issues a $\overline{SELECT}$ signal.
   □ With both $\overline{SELECT}$ and $\overline{IN}$ present at its inputs, the out-
      put of G1 goes HIGH. This becomes our IN signal.
Input operations:
   □ $\overline{IORQ}$ and $\overline{WR}$ go LOW, and the port address appears on
      bits AØ through A7 of the address bus.
   □ The input of the 7442 sees 0010, so the No. 2 output (pin
      3) goes LOW. This is the $\overline{OUT}$ signal.
   □ At this time, $\overline{SELECT}$ goes LOW.
   □ Both inputs of gate G2 are now LOW, so its output goes
      HIGH to produce the OUT signal.
   Figure 17-11 shows a method based on the 7442, but without
the NOR gates, Here we are using the D-input to detect the
$\overline{SELECT}$ output of the address decoder. We still use the codes
0100 and 0010 for $\overline{IN}$ and $\overline{OUT}$, respectively, but they are now
generated without the NOR gates.

244

If a SELECT signal is available, then the output of the address decoder goes HIGH for the correct address. This would generate the codes 1100 and 1010 for IN and OUT, both illegal on the 7442. To over come this slight hitch, we must pass the select signal through an inverter to create a $\overline{\text{SELECT}}$ signal. This is done in Fig. 17-11. See also Fig. 17-12.

If we wanted to use 74154 binary-to-1-of-16 decoder, however, we could do substantially the same thing, but without the inverter. One might wonder, however, if that is economical.

One last approach to creating I/O capability is to use one of the complex I/O function ICs offerred by microprocessor manufacturers. These chips are usually called peripheral interface chips, or something similar (PIO, PIA, etc.). Zilog, Inc., the prime manufacturer of the Z80-family of devices, makes at least two types: serial (Z80-SIO) and parallel (Z80-PIO). Figure 17-13 shows the Z80-PIO parallel interface chip connected to the Z80 CPU chip. Note that certain pinouts on the PIO have the same designations as pinouts on the Z80 chip. This means that they are to be connected to the designated pins on the Z80 for proper operation. The Z80-CPU and Z80-PIO are, then, a matched pair designed to have like pins connected together. In this application we also pair $\overline{\text{IORQ}}$ and $\overline{\text{CE}}$ (chip enable), on the theory that the Z80-PIO need not be enabled unless an input/output operation is taking place, and the $\overline{\text{IORQ}}$ terminal indicates the I/O state. You may wonder why one would have a separate chip enable terminal at all. The reason is that

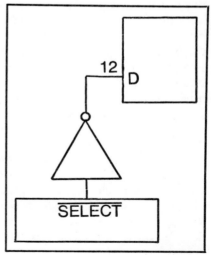

Fig. 17-12. Inverted $\overline{\text{SELECT}}$ to make SELECT signal.

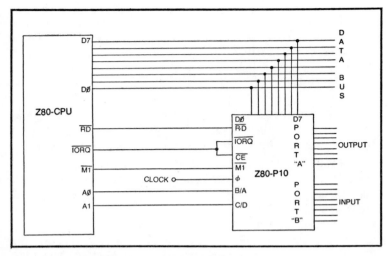

Fig. 17-13. Interfacing I/O using the Z80 PIO.

the PIO may be used in a circuit where there is more than one type of I/O.

INTERFACING MEMORY

One of the first accessories one adds to a computer is external memory. Although some of the latter CPU chips contain on-board memory locations, most require external memory before they can be used to do any real work. At one time, "memory," consisted of tiny doughnut-shaped ("toroidal") cores of ferrite material. These were magnetized to store data. The circuitry required to support core memory was tremendous by today's standards. In modern computers we find solid-state memory, which requires little in the way of support circuitry. Solid-state is vastly easier to support, is a lot cheaper, and much faster operating. Core memory, on the other hand, was nonvolatile; i.e., you don't lose the data just because the power supplies goes off. Many solid-state memories using the popular bipolar transistor technology, unfortunately, *are* volatile; they seem to suffer an, errr, uhhh, errr, a whatacallit, *amnesia* on power down. Recent solid-state memory devices, however, have partially solved that problem by using MOS technology. A small bias voltage, at almost zero current, will sustain memory even though power is off. This bias can be obtained from a small battery. You may have already seen this technology in some consumer devices. Some calculators and electronic checkbooks (a form of

246

calculator) boast continuous memory. These machines remember the last data (your present checking account balance) present in the registers at power turn/off. At least one television receiver uses electronic memory for their programmable TV tuner. The customer would program 10 to 15 registers with the TV channels present in the locale, and thereby custom tunes the set. A small battery is used to keep the data current, even though the set is turned off, or unplugged. CMOS, PMOS and NMOS memory requires so little current that the memory hold batteries are said to last two years!

Read only memory, or ROM, memory chips are preprogrammed (permanently or semipermanently in the case of EPROMs). The CPU cannot write data into a ROM, but can read data *from* memory; hence, the name *read only memory* (ROM).

Figure 17-14 shows how to interface a certain class of ROM device. This case covers those devices with two chip enable inputs. CE1 is active-HIGH and CE2 is active-LOW. Input CE2 is connected to A10 of the address bus. Input CE1, however, requires two conditions that must be met before it can become active: $\overline{RD}$ and $\overline{MREQ}$ must both be LOW, indicating that a memory read operation is being executed by the CPU.

In the simplified case shown, we use bit A10 to turn memory on and off. Here we are designating the 1024 addresses of the ROM as the lowest 1024 bytes of the allowable 65,536 byte memory space. A0 through A9, then, are active and A10 can be used as a chip enable, freeing the rest of the available addresses for random access read/write memory (RAM). Bit A10 of the address bus goes HIGH, disabling the ROM, for those addresses higher than 1024.

An alternate system, for use with ROMs and EPROM that store only 256 bytes (like the 1702) is shown in Fig. 17-15. This

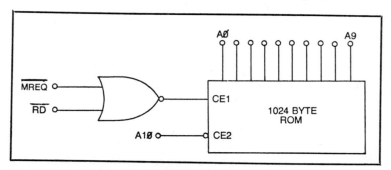

Fig. 17-14. Using MREQ and RD signals to interface ROM.

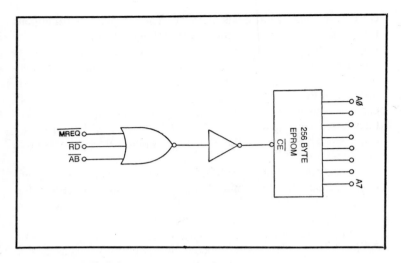

Fig. 17-15. Interfacing the 256-byte ROM.

device uses only eight address bits (AØ through A7) and has a single active-LOW chip enable input.

We want the $\overline{CE}$ to go LOW only when three conditions are met:

☐ $\overline{MREQ}$ (memory request) is LOW.
☐ $\overline{RD}$ (read) is low.
☐ A8 is LOW.

The first of these three ($\overline{MREQ}$ and $\overline{RD}$) are the Z80 control signals that together indicate a read from memory operation is taking place. The third tells us that the address on the address bus is in the lower 256 bytes of memory space. All other possible addresses has bit A8 HIGH.

We can accomplish this job using a three-input NOR gate. Its output stays LOW as long as any *one* input is HIGH. It will go HIGH only when all three inputs are LOW. This happens in Fig. 17-15 only during memory read operations in the lower 256 bytes of memory. The NOR gate ouput must be inverted (it is active-HIGH) before it will properly drive the active-LOW input of the EPROM ($\overline{CE}$).

You may be confused by our seeming use of NOR gates for apparently NAND functions. In the case above, the output is HIGH only for the condition $\overline{MREQ}$ AND $\overline{RD}$ AND $\overline{A8}$. This is merely terminology: a positive logic NOR gate is a negative logic NAND

248

gate (and vice versa). TTL and most CMOS gates are assigned names that assume positive logic. In this circuit, however, we have active-LOW chip enable functions, so the logic is negative. This is indicated by the "NOT bar" symbols over MREQ, RD, and A8.

Figure 17-16 shows an alternate form of the circuit in Fig. 17-15. Here we are not using a triple input NOR gate IC., but instead are making a three-input NOR function by properly interconnecting an inverter to a pair of two-input NOR gates. This does not, at first blush, appear to be good economics; it requires three gates extra. But under the right circumstances, digital equipment makers will opt for this type of circuit because it *is* economical. Keep in mind that typical ICs in both TTL and CMOS are *hex* (6) inverters and quad (4) two-input gates. If there are inverter and NOR gates used elsewhere in the circuit, then there might well be a sufficient number of spare inverter/NOR-gate sections to make the circuit of Fig. 17-16.

Alternatively, what if there is no spare inverter available? We would need an IC with a three-input NOR and another with six inverters. This results in five wasted inverted sections. But a single quad, two-input, NOR gate will suffice for Fig. 17-16 if two of the NOR gate sections were inverter-wired (see two examples in Fig. 17-17).

Thus far, we have considered only the simplest cases; i.e., ROM/EPROM in the lower portion of memory. But as man does not live by bread alone, computers do not live on 1k bytes of ROM alone. In the next section we will consider random access read/write memory to 64k bytes (65,536).

RAM R/W MEMORY ORGANIZATION

Although IC manufacturers burble (or, should that be *bubble*) a lot about 64K byte chips for memory, most cannot as of this time

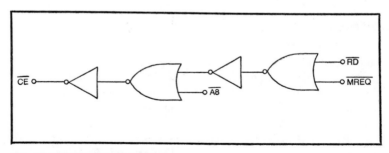

Fig. 17-16. Using discrete gates.

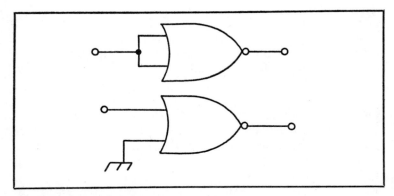

Fig. 17-17. NOR gate inverter connected.

actually deliver them in quantity at good prices. Most computers use arrays or banks of memory ICs, each containing a fraction of the total memory. These are connected with appropriate decoding to the address bus and directly to the data bus.

Consider Fig. 17-18. Here we see the popular 2102 memory ICs, which are 1024X1 bit devices. This means that the address bus can access 1024 1-bit locations. To make a 1024 bank of one byte locations, then, requires eight ICs. A total of 64 devices are required to make a 64K array of memory locations. This circuit is actually somewhat simplified in that only one data bus is shown. The actual 2102 devices has *data to CPU* and *data from CPU* terminals, so a little external logic is required to merge these into one data bus.

Another organization scheme is shown in Fig. 17-19. In this case, two ICs are paired, each able to store 256 four-bit words. Two devices together, then, will store 256 bytes. Also shown in this figure is the $\overline{RD}$ and MREQ/A1Ø logic required to turn on this bank of memory.

MEMORY BANK SELECTION

When memory is organized into blocks or banks, it becomes necessary to be able to select the required bank and the specific address within that bank. Figure 17-20 shows a bank select scheme for eight-bit microcomputers capable of addressing 8K bytes of memory in 1K byte blocks. Once again, we see the use of the 7442 BCD-to-1-of-10 decoder IC

Bits A1Ø through A9 of the address bus are sufficient to address 1024 locations in memory. The pins for these bits are all

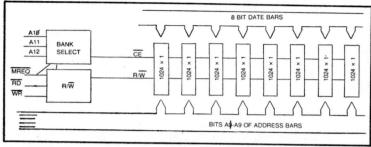

Fig. 17-18. Interfacing to RAM organized in 1024X1 bit chips.

tied in parallel such that all AØ pins are tied together, all A1 pins are tied together and so on, forming an address bus. But unfortunately, there is not much use in addressing the same location in all eight blocks simultaneously. We need a method for determining which

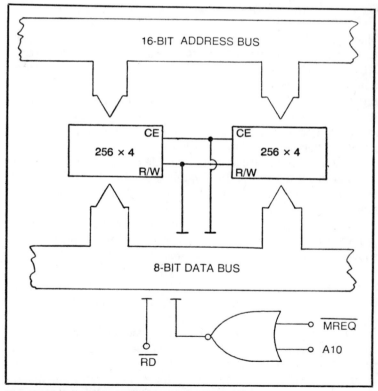

Fig. 17-19. Interfacing RAM organized in 256×4 chips.

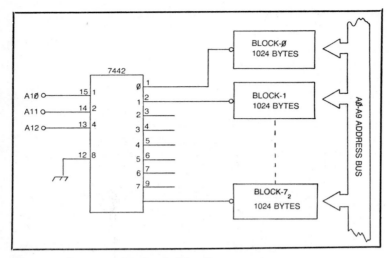

Fig. 17-20. Generating block select function.

block of 1024 bytes is intended, and then to turn on the CE terminals of those chips. We start by examining bits A10 through A13 (see Fig. 17-21). Note that bit A13 is always LOW, so it can be ignored. If we assign the eight blocks the number 0 through 7, we find that bits A0 through A12 form a binary number equivalent to the block number. We may now use A10 through A12 as the "1-2-4" inputs to a 7442. The "8" input of the 7442 (pin No. 12) can be grounded because A13 is always zero.

We now have a circuit that is capable of selecting the correct location in the correct block for the applied address. The 7442 outputs are active-LOW, so will interface directly with most common memory IC devices.

For computers that allow addresssing of more than 8K bytes of memory, we must extend this concept a little. Figure 17-21 shows a circuit using eight 8K byte banks (each like Fig. 17-20) to provide a complete 64K memory. Each eight kilobyte bank has its own 7442 to sense the state of bits A10 through A12. This scheme is identical to Fig. 17-14A. Similarly, bits A0 through A9 of all memory devices are bussed together.

The 64K circuit of Fig. 17-21 uses one additional 7442 decoder to examine bits A13 through A15. The "8" input of the bank select 7442 is grounded permanently, but the "8" inputs of all eight-block select 7442 are used as chip enable pins. Remember, since only eight possibilities (0 through 7) are allowed, the "8"

252

Table 17-1. Condition Codes at Input of 7442.

		A13	A12	A11	A1Ø
Block-Ø	Ø-1K	Ø	Ø	Ø	Ø
Block 1	1K-2K	Ø	Ø	Ø	1
Block 2	2K-3K	Ø	Ø	1	Ø
Block 3	3K-4K	Ø	Ø	1	1
Block 4	4K-5K	Ø	1	Ø	Ø
Block 5	5K-6K	Ø	1	Ø	1
Block 6	6K-7K	Ø	1	1	Ø
Block 7	7K-8K	Ø	1	1	1

input is LOW for all valid addresses. The block select 7442s, then, have their "8" inputs connected to the appropriate outputs (0 through 7) of the bank select 7442.

Let us describe the conditions in the circuit of Fig. 17-21 if the desired memory address is 11613_{10}. This address, written in binary form, is 0010110101011101_{2}. This address can be divided up into location (bits AØ through A9), block (A1Ø through A12) and bank (A13 through A15):

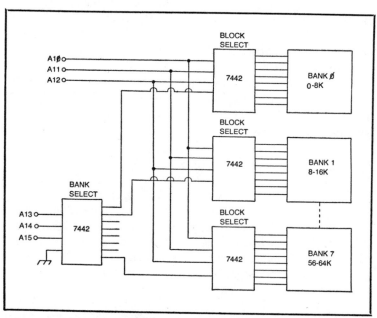

Fig. 17-21. Generating bank and block select.

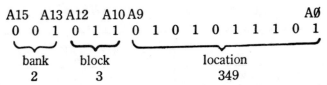

A15 A13 A12 A10 A9   AØ
0 0 1 0 1 1 0 1 0 1 0 1 1 1 0 1

bank   block   location
  2   3   349

We would expect this binary word to address location 349 (out of the possible 1024), in block 3 (out of 8) of bank 1 (of 8). This is at least "in the ball park" because bank 1 is 8 -16K, and 11613_{10} falls within that range. We can check this further, however, by doing a little arithmetic:

$$
\begin{array}{ll}
0101011101 & 349 \\
011\text{-------------} & 3072 \\
001\text{---------------- - - -} & 8192 \\
\hline
0010110101011101 & 11613
\end{array}
$$

When 0010110101011101 appears on the address bus, location 349 in all 64 memory ICs are addressed simultaneously. The proper combination of block and bank select logic will turn on only the single set of eight ICs that contains location 11613. In this case, block No. 3 is selected, so all eight-block select 7442s see 011 on bits A1Ø through A12. But only the block select chip serving bank No. 1 will see the *correct* code 0011; all others see 1011. This is because output No. 1 of the bank select 7442 is LOW, while the others are HIGH; hence, we turn on bank No. 1, block No. 3, and location 349 to select location 11613.

Note that this is only one scheme of several that are commonly used in microcomputers. But this book must, because of space limitations, be representative rather than comprehensive.

MEMORY-MAPPED I/O

Using the regular input-output functions of any computer is sometimes more limited, and time consuming, than the programmer might wish. All I/O instructions in most computers, for example, must be processed through the accumulator. In order to transfer data from a register (usually, but not always, the accumulator) to I/O, one must first pass through the accumulator, and then to the I/O port. Memory instructions, on the other hand, are often arranged to allow direct transfer between the accumulator and a location, another register and memory location, or from one memory location to another. These avoid the step of first loading the data into the accumulator before writing it to an I/O port. Also, another limiting factor is that there are only 256 different possible I/O locations, and 65,536 memory locations.

Memory-mapped I/O is a technique that allows us to treat peripherals and I/O ports as memory locations. This assumes, of course, that unassigned memory exists. Most microcomputers use less than 40K, and many less than 32K, of the available 64K. In those machines, it is a simple matter to add memory mapped I/O. If there is a full complement of memory (64K), then we can still use use memory mapping only we must sacrifice the location to the peripheral. This schemes requires us to remember to not program, or write data into, that location. Any data written to this location and to the peripheral assigned to that location.

Figure 17-22 shows a typical memory-mapped peripheral device. In this case, it is a digital-to-analog converter, or DAC (see Chapter 18). A DAC will produce an output voltage or current that is proportional to the binary word applied to its input bits. We may, therefore, use a DAC to produce an output voltage for an oscilloscope, chart recorder, or analog control circuit under control of the computer. The binary word applied to the DAC is held in the eight-bit 74100 data latch. Data from the data bus is transferred to the latch outputs when line D is brought HIGH. The decoding scheme shown here is similar to other memory operations (again the Z80 uP chip is used as an example). Coexistence of $\overline{MREQ}$ and $\overline{WR}$ causes point A to go HIGH, which is inverted to force point B LOW. Point C, the remaining input of the NOR gate G2, goes LOW only when the memory address assigned to the DAC is present on the address bus. In any appropriate operation, where it is desired to send an output word to the DAC, both C and B are LOW, forcing

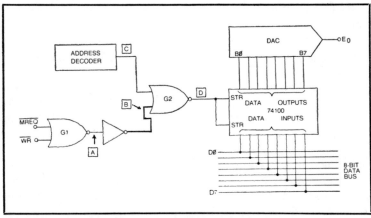

Fig. 17-22. Memory-mapped I/O (in this example using a DAC).

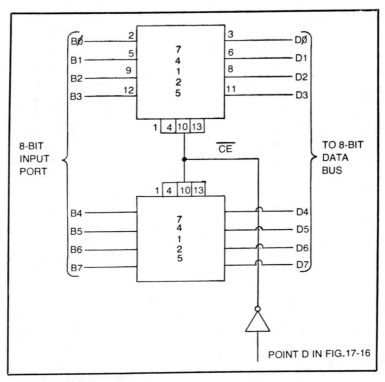

Fig. 17-23. Tri-state input port.

D (the *strobe* terminal of the 74100 data latch) HIGH. Data on the DØ - D7 bus at that time instant is transferred to the latch output and held.

We may also use this same circuit, minus the DAC of course, as an output port. The outputs of the 74100 will then appear to the outside world as a TTL-compatible, eight-bit, binary output port.

An input port can be formed by using a tri-state bus driver in place of the 74100, and the $\overline{RD}$ signal instead of the $\overline{WR}$ signal. An extra inverter is also needed if the device uses an active-LOW chip enable (the usual case). Examples of typical circuits are shown in Figs. 17-23 and 17-24.

The circuit is Fig. 17-23 uses two quad buffers that have tri-state outputs (TTL device 74125). A "tri-state" device has three possible output states instead of the two normally associated with binary devices:

□ Low impedance to ground (LOW).

256

☐ Low impedance to V+ (HIGH).

☐ High impedance to both ground and V+ (the output looks like a high-value resistor to the bus).

Tri-state devices can be connected to the data bus, and will not load the circuit unless the chip enable terminals (1, 4, 10, and 13 in the 74125 device) are LOW. If $\overline{CE}$ is LOW, then, the DØ through D7 outputs connected to the data bus will follow the BØ through B7 inputs. An inverter is required on the $\overline{CE}$ line to make it compatible with the circuit of Fig. 17-22. Of course, there might also be alternate logic schemes used that already provide an active-LOW chip enable function.

The above circuit uses the TTL 74125 device. Similar 2×4 bit circuits may also be used. One that comes immediately to mind is the Intel 8216 device. It is intended for the 8080 microprocessor but works equally well in other applications.

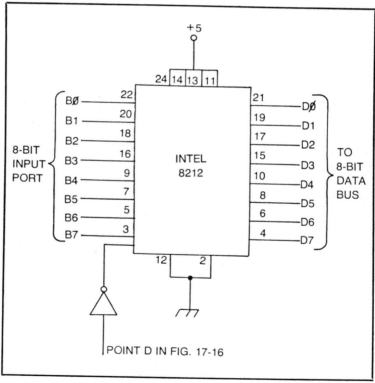

Fig. 17-24. Using Intel 8212 tri-state I/O device.

257

An 8×1 bit circuit, capable of inputing an entire byte with one chip, is shown in Fig. 17-24. It uses the Intel 8212 device. This chip is also intended for service with the 8080 microprocessor but is often found in other applications. Like the circuit of Fig. 17-23, it is not a latch, but a tri-state buffer. When point D goes HIGH, the port is enabled. Data present at that instant on BØ through B7 is transferred to DØ through D7. When point D is LOW again, after the memory read command is executed, the outputs again go into the high-impedance state. Under that circumstance, the data bus doesn't even see the outputs of the 8212.

Both circuits of Figs. 17-23 and 17-24 can be used as an ordinary input port. We may also use the 74100 configuration of Fig. 17-22 as an output port. Modifications would be needed, but they are slight: replace $\overline{\text{MREQ}}$ with $\overline{\text{IORQ}}$ and use only AØ through A7 of the address bus.

Chapter 18
Data Conversion

We know that digital electronic circuits are happy only with *binary* signals. This means that the signals can take on only one of two possible values designated HIGH or LOW (or "1" and "0"). In-between values will either be ignored, or will be accepted as one of the two permissable values. In TTL circuits, for example, HIGH is defined as +5 volts and LOW is 0V (grounded). But in actuality, "0" means anything in the 0V to 0.8V range, and "+5" means "greater than +2.4V". The ranges, then, are quite wide and analog voltages that fall within either range will be treated as if it were a real binary signal.

In Chapter 1, we discussed several different types of signal, of which digital is but one. You learned that analog signals can be continuously variable over a given range and domain. Such a signal can take on any value at any time. The sampled signal, on the other hand, can take on any value, but only at certain discrete times. Neither of these signals are compatible with digital circuits. The true digital signal of more than one bit can take on only values of the binary number system, and often is also time constricted.

Unfortunately, much electronic instrumentation produces either analog signals or sampled signals as the output. If these signals are to be processed in a digital computer, or using digital circuitry a *data converter* is required to change input values to a form that can be recognized by the digital circuitry; i.e., it must be converted to binary words.

There are actually two classes of data converters: *analog-to-digital converters* (ADC) and *digital-to-analog converters* (DAC). The ADC converts the analog voltage or current input to a binary word, as described above. The DAC performs the inverse process;

it converts binary words to a proportional voltage or current output. The DAC may be used to drive analog display instruments, such as the oscilloscope or strip-chart recorder, or to drive analog control circuitry.

Transducers are devices that convert physical parameters such as force, pressure, velocity, position, temperature, light intensity, into an electrical signal. Most common transducers output a signal in the form of a voltage or current that is somehow proportional to the applied stimulus parameter. One common temperature transducer, for example, produces an output voltage proportional to the applied temperature. The scaling factor is given as 10 mV/°K. The freezing point of water is 273° K (0 °C). Note that the size of the Kelvin degree is the same as the Celcius degree, so convert °K to °C by subtracting 273). At freezing, then, the output voltage would be

$$273 \text{ °K} \times \frac{10 \text{ mV}}{\text{°K}} = 2.73 \text{ volts}$$

But before we can apply the data from a transducer or any electronic amplifiers and other circuits connected to the transducer, to a digital instrument, the output voltage (or current) must *first* be converted to a binary word. The data converter is used to accomplish this neat trick.

Similarly, some signals already exist in the form of voltages or currents. In medical electronics, for example, they use electrodes on the patient's body to pick up minute biopotentials (electrical potentials created by living cells). See TAB book No. 930, *Servicing Medical & Bioelectronic Equipment*. One commonly sought biopotential is the electrocardiogram, or ECG (also called EKG after the German spelling). This electrical signal is produced by the beating heart. Before any ECG monitoring or diagnostic computers can do their work, however, the analog biopotentials must be passed through a data converter to produce a series of binary words that represent signal.

Our last example, which will be the first class of data converters that we consider, is the DAC. There are two places where we would want to convert a digital signal—binary word—to an analog signal: display and control. In the case of the display, we might want to let human operators see the waveform on an oscilloscope screen, strip-chart recorder, X-Y recorder, or even an ordinary voltmeter, although there are non-analog methods the same job can be done. In the case of the control system, we might use an ADC to supply input data, and then output a computed result to a DAC to change or modify the system somehow.

Neither ADC nor DAC can represent all possible values or states of the analog signal. Remember, an analog signal can take on an infinite number of values within an assigned range (0V to 10V). The data converter, on the other hand, is limited to discrete amplitude values and discrete times. The former depends upon the range and bit-length (6-bits, 8-bits, etc.), while the latter depends upon the *conversion time*, or the time required to make a complete, valid, data conversion. Say we have an 8-bit ADC that is operated over an input voltage range of 0 to 10 volts. If eight bits are available, then there are 2^8, or 256, different binary states possible in the output word. All possible analog values must be shoehorned into 256 different states. The zero volts state is usually represented by the binary word 00000000. There are, then, 255 remaining combinations to represent all other analog values in the range. Clearly, then, only certain discrete input voltage values can be represented exactly. All others will contain a small error term. This leads us to the topic of converter *resolution*.

DATA CONVERTER RESOLUTION

The *resolution* of a ADCs and DACs can be divided into three ranges: 2 to 6 bits, 6 to 10 bits, and over-10-bits. Those converters in the under 6-bits range are generally capable of accuracy in the 1 to 2 percent range, while those in the 6 to 10 bits range produce accuracies on the order of 0.1 to 1 percent. Even greater accuracy is possible in the over-10-bits class, but here some degree of caution is required. It is very easy to design a converter with 10 to 16 bit word lengths, but many have accuracies similar to 8-bit designs Since we are always comparing an unknown analog voltage to a known reference source, the accuracy of the reference voltage (current) becomes important. The design of an 8-bit voltage reference supply is almost trivial, but the problem becomes more acute in over-10-bit designs.

Additionally, other problems tend to deteriorate the accuracy of the over-10 bit data converter. The gain error of any input amplifiers, for example, can easily become significant in this range. Another error is the hysteresis range of the voltage comparator circuit that is used to compare the reference and the input signal. The hysteresis is the minimum signal difference applied between the inputs of the comparator that will cause a change in the output state. This value ranged from 4 mV to 23 mV in a group of randomly selected LM311 comparators tested by the author.

Perhaps an example would best serve to illustrate this problem. In all ADCs, the noise should be less than one least-significant-bit (1-LSB). Assume an ADC designed to operate over the 0V to 10V range of input signals. The minimum resolution is always given by:

$$E_{LSB} = \frac{10 \text{ volts}}{2^n}$$

where E_{LSB} is the voltage change required to change the LSB of the data converter, and n is the converter bit-length.

For an 8-bit ADC, then:

$$E_{LSB} = \frac{10 \text{ volts}}{2^8}$$

$$E_{LSB} = \frac{10 \text{ volts}}{256}$$

$$E_{LSB} = 0.04 \text{ volts} = 40 \text{ mV}$$

For a 12-bit ADC:

$$E_{LSB} = 10 \text{ V}/2^{12}$$

$$E_{LSB} = 10 \text{ V}/4096$$

$$E_{LSB} = 2 \text{ mV}$$

In the 8-bit case, the 40 mV LSB value is clearly greater than the 23 mV worst-case comparator hysteresis, so there is little cause for concern. But in the 12-bit case, the hysteresis of the comparator can be eleven times greater than the 1-LSB voltage. In this example, the noise bit is (LSB+3), so our "12-bit" ADC will perform no better than an 8-bit ADC! It is generally conceded by ADC designers than words like "simple" and "low cost" form an exclusive set that does not include "high precision." One can, however, make use of low-cost or simple ADCs by properly recognizing true requirements. Over specifying ADCs can, indeed, be expensive.

When designers specify data converters they need to know the bandwidth of the input signal that will faithfully reproduce the waveform. All signals can be represented by a Fourier series of sine and cosine waves. Recall that a square wave is made up from a fundamental and an "infinite" number of harmonics. The same is true of all other Waveforms: Only the pure sine wave contains no harmonics. The bandwidth required to faithfully reproduce or process an analog signal is approximately the frequency of the highest harmonic that can cause a recognizable change in the waveform if it were missing. For the human ECG described earlier, for example, it has been found that a bandwidth of 100 Hz is required. This means that amplifiers with a 100-Hz frequency response must be used to process the analog signal. It also sets the required *sampling rate* of the data converter. There is a theorem followed by circuit designers that demands a minimum sampling rate that is *twice* the highest frequency component present in the signal. Our 100-Hz ECG waveform, then, must be sampled at a rate of not less than 200 samples per second or there will be errors. In actual practice, faster rates are generally used. The data converter used, then, must be capable of making at least 200 conversions per second, or the waveform may be seriously compromised.

DAC CIRCUITS

It is generally best to consider the different types of DAC circuit prior to discussing ADC techniques. The reason for this seemingly illogical procedure is that DACs are used as an integral part of several popular ADC circuits.

The purpose of the DAC is to create an output voltage or current that is proportional to the product of a reference source and a binary input word. Considering only (for now) the voltage case:

$$E_o = \frac{A\,E_{ref}}{2^n}$$

Where E_o is the output voltage, E_{ref} is the reference voltage in the same units of E_o, A is the binary word applied to the DAC input (decimal expression), and n is the bit-length of the DAC input ($n = 8$ for 8-bit DACs).

There are two principal forms of DAC circuit, and both are based upon simple resistor networks. One of these is the *binary weighted resistor ladder*, and the other is the *R-2R resistor ladder*.

Figure 18-1 shows the binary weighted resistor ladder. The ladder output is connected to the inverting input of the operational amplifier, which is also a summing junction. If you are unfamiliar with operational amplifier theory, let me recommend TAB Book No. 787, *Op-Amp Circuit Design & Applications*. The resistors in the ladder are said to be binary weighted, because their values are related to each other by powers of two. If the lowest value resistor in the circuit is taken as R, then the next higher value will be 2R, followed by 4R, 8R, 16R, 32R, etc., up to the nth resistor in the ladder which has a value of $(2^{n-1})R$.

In this circuit we show discrete switches used to turn on the resistors. These correspond to the digital inputs actually used in real DACs, which, incidentally, take on the form of electronic switches, so the analogy holds true. Switches B1 through B_n represent the binary word that is applied to the input of the DAC. Switches are used to connect each input to either ground (logical-0) or some reference voltage (logical-1). Switches B1 through B_n allow currents I1 through I_n to flow when they are closed, respectively.

Ohm's law tells us that each current is equal to the quotient of E and the value of the resistor. For example,

$$I1 = E/R$$
$$I2 = E/2R$$
$$I3 = E/4R$$
$$I4 = E/8R$$

$$I_n = E/(2^{n-1})R$$

The total current flowing into the summation junction of the op-amp is:

$$I_A = \sum_{i=1}^{n} \frac{A_i E}{2^{(i-1)R}}$$

where I_A is the current into the op amp summing junction from the resistor ladder network, E is the reference voltage in volts (V), and R is the resistance of R1 in ohms.

Note that the Σ symbol means "summation," and will be used elsewhere in this book. What it means, simply, is that we replace i with an integer that is one higher on each go-around, until n is reached. For example, consider a 3-bit converter (n=3):

$$I_A = \sum_{i=1}^{3} \frac{A_i E}{2^{(1-1)}R} =$$

$$\frac{A_1 E}{2^{1-1}R} + \frac{A_2 E}{2^{2-1}R} + \frac{A_3 E}{2^{3-1}R}$$

$$I_A = \frac{E}{R}\left[\frac{A_1}{2^0} + \frac{A_2}{2^1} + \frac{A_3}{2^2}\right]$$

$$I_A = \frac{E}{R}\left[\frac{A_1}{1} + \frac{A_2}{2} + \frac{A_3}{4}\right]$$

where A_1, A_2, and A_3 are the bits at that input, and may be either 1 or 0. Note that in the previous equation, A_i will always be either 1 or 0; i.e., the switch is either turned on or turned off. Now, let's consider the operation of the circuit.

Operational amplifier theory tells us that

$$I_A = -I_f$$

and

$$E_o = I_f R_f$$

We can substitute the first equation into the second to obtain

$$E_o = -I_A R_f$$

We can now use this result to get:

$$E_o = -R_f \sum_{i=1}^{n} \frac{A_i E}{2^{(i-1)} R}$$

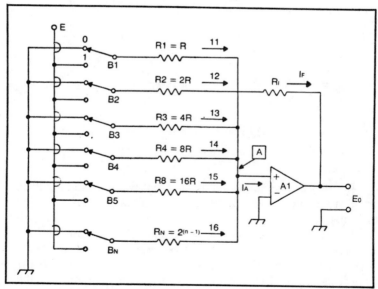

Fig. 18-1. Binary-weighted resistor ladder DAC.

But E and R are constants, so we usually write them outside of the "summation" expression:

$$E_o = \frac{-ER_f}{R} \sum_{i=1}^{n} \frac{A_i}{2^{(i-1)}}$$

☐ **Example:**
A four-bit DAC using the binary weighted resistor ladder technique uses a +2.56 V reference source. The value of R_f is 10 kohms, while R is 3.3K ohms. Find the output voltage for the binary input word 1101. (Hint: $A_1 = 1$, $A_2 = 1$, $A_3 = 0$, and $A_4 = 1$).

$$E_0 = \frac{-ER_f}{R} \sum_{i=1}^{4} \frac{A_i}{2^{(i-1)}} \qquad E_o = \frac{(-2.56)(10k)}{(3.3\ k)}$$

$$\left\{ \frac{A_1}{2^{(1-1)}} + \frac{A_2}{2^{(2-1)}} + \frac{A_3}{2^{(3-1)}} + \frac{A_3}{2^{(3-1)}} + \frac{A_4}{2^{(4-1)}} \right\}$$

$$E_o = (-7.76\ V) \left\{ \frac{1}{2^0} + \frac{1}{2^1} + \frac{0}{2^2} + \frac{1}{2^3} \right\}$$

$$E_o = (-7.76\ V) \left\{ \frac{1}{1} + \frac{1}{2} + \frac{0}{4} + \frac{1}{8} \right\}$$

$$E_o = (-7.76\ V) \quad \{ 1 + 0.5 + 0 + 0.125 \}$$

$$E_o = (-7.76\ V) \quad (1.625) = -12.61V$$

It is not always immediately apparent when examining ideal circuits presented in a textbook, but there are problems in trying to implement this circuit in actual practice. Bertrand Russell is attributed with the following quote: "There must be an ideal world, a sort of mathematician's paradise, *where everything works as it does in textbooks.*" The problem with this circuit is that the values of the input resistors tend to become very large and very small at the ends of the network. This becomes even more acute as the bit length of the converter becomes longer. There is little trouble when the bit length is less than eight-bits, but at longer lengths resistor values become ridiculous. Keep in mind that the highest resistor in an eight-bit converter will be 128 times the value of the lowest resistor. Yet the value of the lowest resistor must be such that E/R does not produce a large current that could burn out the resistor (or the op amp, for that matter). We also find that most easily obtained operational amplifiers cannot operate with the finds of input currents that will be found in the 128R branch of the network. Consider the case above, where a 2.56V reference source was used. If R=1K ohm, then 128R is 128,000 ohms. Current I8 will be 2.56/128000 = 0.000020 amperes, or 20 μA. Most nonpremium grade operational amplifiers will not easily resolve this current. DACs built with these parameters might tend to "drop bits" in the least significant (I8) position. Additionally, have you ever tried to buy resistors in some of the odd "power of two" multiples of any reasonable R-value? They are not too easy to obtain on short notice and will cost a premium price when available. As a result, we rarely find the binary weighted ladder in real applications, and then only in 3 to 6 bit applications.

The R-2R ladder DAC circuit is considered superior to the weighted ladder in several respects. One is the fact that all of the

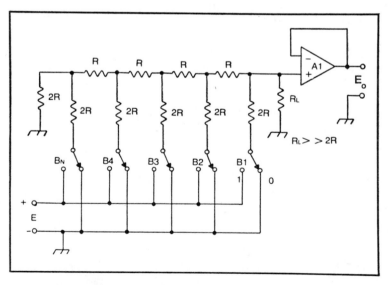

Fig. 18-2. R-2R resistor ladder DAC.

resistors will have one of two values, R or 2R. This simplifies production of the unit, and results in a more consistent precision without a lot of trimming. Most commercial DACs use the R-2R resistor ladder instead of the weighted ladder.

The output voltage of the resistor ladder shown in Fig. 18-2 is applied to the noninverting input of an operational amplifier. In this case, we are using the high input impedance of the op amp to prevent loading of the network, and to provide buffering to the outside world. We could just as easily obtain some gain, but that is not too common in commercial IC DACs. Assuming that the value of R_L is very much higher than R or 2R, we can write the output voltage expression as:

$$E_o = E \sum \frac{A_i}{2^i}$$

(The terms are defined similarly to the earlier expression given for the weighted ladder DAC).

☐ **Example:**

A R-2R four-bit ladder DAC has a 2.56 V reference source and sees a binary word of 1101 at the input. Calculate the output voltage.

$$E_o = (2.56 \text{ V}) \sum_{i=1}^{4} \frac{A_i}{2^i}$$

$$= (2.56 \text{ V}) \left\{ \frac{1}{2^1} + \frac{1}{2^2} + \frac{0}{2^3} + \frac{1}{2^4} \right\}$$

$$= (2.56 \text{ V}) \left\{ \frac{1}{2} + \frac{1}{4} + \frac{0}{8} + \frac{1}{16} \right\}$$

$$= (2.56 \text{ V}) (0.5 + 0.25 + 0 + 0.0625)$$

$$= (2.56 \text{ V}) (0.8125) = 2.08 \text{V}$$

Full-scale output from the R-2R resistor ladder is given by the expression:

$$E_{o \text{ (max)}} = \frac{E (2^n - 1)}{2^n}$$

where $E_{o \text{ (max)}}$ is the maximum value of the output voltage E_o, E is the reference voltage, and n is the bit-length of the DAC digital input word.

□ Example:

Find the maximum output voltage of an eight-bit DAC that uses a 2.5 VDC reference supply.

$$E_{o (max)} = \frac{(2.5 \text{ V}) (2^8 - 1)}{2^8}$$

$$= \frac{(2.5 \text{ V}) (256 - 1)}{(256)}$$

$$= \frac{(2.5 \text{ V}) (255)}{(256)}$$

$$= 2.49\text{V}$$

The input to a DAC is a binary word; it can exist only in those discrete states defined by the binary number system. The output of a DAC is dependent, in part, upon the digital input word. The DAC output voltage, therefore, can exist only with certain discrete values. Each successive binary number changes the output an amount equal to the change created by the least significant bit (LSB). This voltage is often called the 1-LSB voltage, and is expressed by:

$$\Delta E_0 = \frac{E}{2^n}$$

Consider an eight-bit DAC with a +10 VDC reference source:

$$\Delta E_o = \frac{(10 \text{ V})}{2^8}$$

$$= \frac{(10 \text{ V})}{(256)}$$

$$= 40 \text{ mV}$$

The 1-LSB voltage, ΔE_o, is the smallest change in output voltage that can occur. In most cases, if we let the zero output state be produced by the binary word 00000000, the maximum output voltage will be E- ΔE_o.

ANALOG-TO-DIGITAL CONVERSION

Analog-to-digital conversion (ADC) is the inverse process to digital-to-analog conversion. The ADC takes an analog voltage or

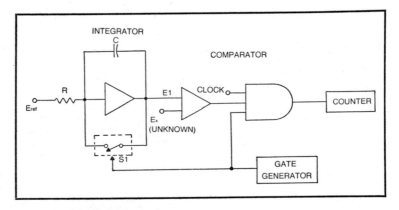

Fig. 18-3. Single-slope integrator.

current input and produces a binary word at the output proportional to the analog input value.

Although many different analog-to-digital converter (ADC) schemes have been designed over the years, only a few have become available in the low-cost form of integrated monolithic or hybrid circuits. This chapter will consider only those that are readily available in that form. These include *single-slope integration, dual-slope integration, voltage-to-frequency conversion, ramp* (or *servo) flash,* and *successive approximation.* We will consider these techniques, and discuss their respective advantages or disadvantages.

Single-Slope Integration

The two integration techniques are examples of *indirect* analog-to-digital conversion. The analog signal is first converted to a voltage function of time, which is then converted to a binary, or BCD, number. In both designs, the principal components are an operational amplifier integrator, a comparator, gating logic, and a digital counter. The counter might have either binary or BCD outputs depending upon the application. In digital voltmeters (which ordinarily use dual-slope integration), for example, the counter would be BCD so that its output may be displayed on LED or liquid crystal decimal readouts. In a microcomputer system, on the other hand, either binary, or the special version of binary called twos' complement is used instead of BCD.

Figure 18-3 shows the circuit for a single-slope integrator ADC, while the timing diagram is shown in Fig. 18-14. During

270

periods when no conversion is taking place, a switch keeps the integrator from accumulating a charge. But when a gate command is received, the integrator output voltage (E1) begins to ramp upwards, and the binary counter begins to accumulate clock pulses. When E1=E_x, the comparator output drops LOW, and thus shuts off the flow of clock pulses into the counter. The output state of the counter at this time, then, is proportional to input voltage E_x.

Single-slope integration is easy to implement, but in its simpler forms suffers much from several inadequacies; the integrator output ramp may be nonlinear, or have the wrong gain (1/RC). This type of converter is sensitive to changes in clock frequency that occur during the conversion period. Although precision ADCs based on single-slope integration have been built, their added cost is sufficient to make other designs more attractive. The single-slope circuit is used primarily in very low cost digital voltmeters.

Dual-Slope Integration

Figure 18-5 shows the block diagram for a dual-slope integration ADC, while the timing diagram is shown in Fig. 18-6.

As in the case of the single-slope circuit, the principal components are an operational amplifier integrator, voltage comparator,

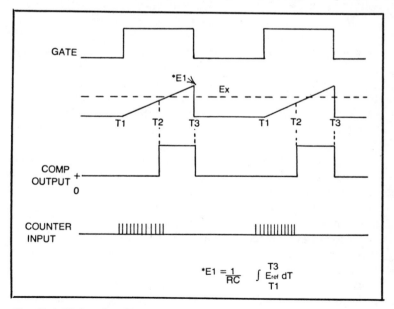

Fig. 18-4. Timing diagram.

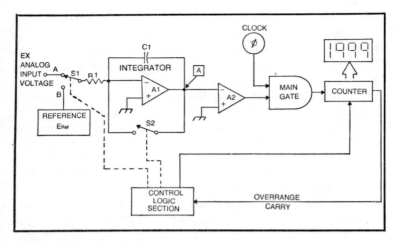

Fig. 18-5. Dual-slope integrator.

and a digital counter. The comparator output will remain LOW if the integrator output is zero, and will be HIGH if there is more than a few millivolts of output from the integrator.

At the beginning of the conversion period, the control logic will momentarily short the integrator capacitor by closing switch S2. It will also insure that switch S1 is set to position A.

When switch S1 is in position "A", the integrator input is connected to the unknown input voltage source, and this causes the integrator output voltage to begin rising. This is shown at time t_{-0} in Fig. 18-6. As soon as E_A is more than a few millivolts greater than zero, the comparator output snaps HIGH. This will turn on the counter gate, allowing pulses to enter the counter circuit. The counter is allowed to overflow (in distinct contrast to the operation of the single-slope circuit!). The output-carry pulse from the counter tells the control logic section that the initial measurement is completed, and that S1 should be switched to position "B". The charge in the integrator capacitor at this time is a voltage function of time, and is proportional to the input voltage. With switch S1 in position "b," we find that the integrator input is now connected to a precision reference source. The polarity of the reference is such that it will tend to *discharge* the integrator capacitor. Since this reference is constant, the integrator capacitor will discharge at a constant, linear, rate. The counter has, by this time, passed through 0000 (on the carry-output it went to 0000) and has continued to accumulate pulses. It will continue to count until E_A is back down to zero. The value of E_A at time t_1 was proportional to E_x. At the same time, the counter display was zero (0000). Since the counter continues to increment as the integrator discharges, the counter state at the instant the gate is closed will be numerically the same as the voltage applied to the input.

The dual-slope integrator is a slow circuit, one of the slowest types of ADC circuit. But it offers several advantages that make it useful in digital voltmeter circuits: relative immunity to noise pulses, relative immunity to clock frequency changes, and immunity to long term clock frequency drift. It's 10 - 40 mS conversion time is reasonable.

One major application for the dual-slope integrator is in digital voltmeters and multimeters (DVMs and DMMs). In that application, the slow conversion time of the ADC is not a disadvantage. In fact, it can be quite an advantage because it tends to average out certain types of noise. In many cases, we have a considerable amount of 60-Hz interference, but the dual-slope integrator DVM will average these to a very low value, provided that the conversion period is greater than 1/60 second, or approximately 17 ms.

Ramp or Servo ADC Circuits

The servo ADC is also known by the names ramp ADC and binary counter ADC. The latter name is derived from the fact that

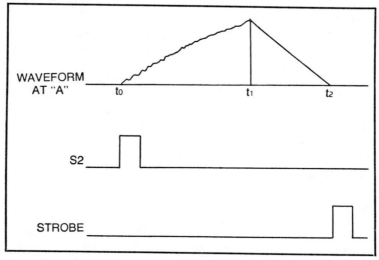

Fig. 18-6. Timing diagram.

273

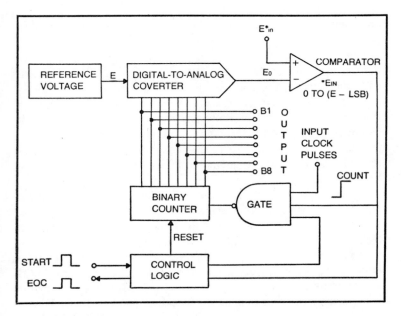

Fig. 18-7. Servo, or ramp, A/D converter.

the circuit uses a DAC with its binary inputs driven by the outputs of a binary counter. Figures 18-7 and 18-8 shows the circuit to a servo ADC. The principal elements are DAC, counter, comparator, gate, and some required control logic. The reference potential source may or may not be present depending upon the nature of the DAC. The comparator is used to keep the gate open, allowing clock pulses to flow into the counter, as long as $E_o \neq E_{in}$.

A start *pulse* initiates the conversion cycle by causing the counter to reset to 00000000, and opening the gate to permit clock pulses to enter the counter. The counter will, therefore, begin to increment, thereby causing the DAC output voltage E_o to begin rising. Figure 18-8 shows this action graphically. The DAC output will continue to ramp upwards until $E_o = E_{in}$. When this equality exists, then the output of the comparator drops LOW, turning off the gate. The binary number at the input of the DAC at this time is proportional to E_{in}.

The control logic section also senses the change in the comparator output level, and uses this indication to generate an output called the *end-of-conversion* (EOC) pulse. This pulse is critical to external circuitry, because there may be data on the output lines all of the time. The circuit cannot tell when the data is valid without an EOC pulse.

The conversion time of the servo ADC is variable, and depends upon two factors: the bit-length and the input voltage. While the bit-length of the DAC/counter (which sets the bit-length of the converter) is fixed for any given circuit, the input voltage is not. The longest conversion time exists when the input voltage is full scale. The conversion time for full scale is 2^n clock pulses (where n is the bit-length). An eight-bit (n=8 servo ADC, then, would require 2^8, or 256, clock pulses in order to make a full scale conversion.

Successive Approximation ADC Circuits

The successive approximation (SA) ADC circuit is somewhat more complex than the simple servo type, even though both are based on a DAC feedback loop. It is used in applications where the long conversion time of the servo ADC is not tolerable; the SA ADC is a lot faster for any given clock speed. For the SA circuit, we find that the conversion time is (n+1) clock pulses. Consider a 10-bit ADC circuit in both servo and SA configurations. The values of the conversion times are:

Servo $\quad T_c = 2^{10} = 1024$ clock pulses

SA $T_c \quad = (10 + 1) = 11$ clock pulses.

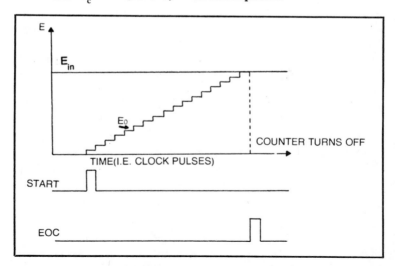

Fig. 18-8. Typical timing diagram.

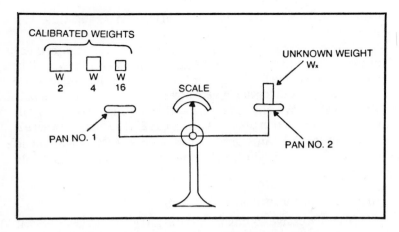

Fig. 18-9. Analogy to the successive approximation A/D converter.

From these, we can see that the SA ADC is 93 times faster than the servo for a full scale conversion. In applications where the conversion time is critical, the SA is clearly superior, despite its complexity.

How Does the SA Circuit Work?

The operation of the SA ADC is a little difficult to understand if we simply describe the electrical operation of the device. But with the use of the analogy of Fig. 18-9, we can more easily see the basic concept. This illustration shows a simple platform balance. The pointer will indicate balance (zero-center) only when the weight on the lefthand side is equal to the weight on the righthand side. In this example, let us stipulate that a weight W will deflect the scale exactly full-scale when pan No. 2 is totally empty. We have a set of calibrated test weights that have weights of $W/2$, $W/4$, $W/8$, $W/16$, etc (to whatever resolution, i.e. bit-length, that we require). If an unknown weight W_x is applied to pan No. 2, the scale will deflect to the right an amount proportional to W_x. To make our precise measurement, we start with $W/2$, and place it on pan No. 1. Three possible situations now exist:

- ☐ $W/2$ equals W_x, in which case the scale is zeroed.
- ☐ $W/2$ is greater than W_x (scale deflected to the left).
- ☐ $W/2$ is less than W_x (scale deflected to the right).

Of course, in situation No. 1, the measurement is completed, and no additional trials are necessary. But let us assume for the moment that one of the other situations exists. In case No. 2, W/2 is greater than the unknown weight, so we must remove W/2 and try again using W/4. This procedure is continued until we obtain situations No. 1 or 3.

In case No. 3, we find that W/2 was less than W_x. We now add more weights in descending succession of value until the scale balances, or the combined weights are greater than W_x. Of course, when the combined weights are equal to the unknown, then the measurement is over and the precise value known. But what about the situation where the value is between two of the possible combinations of weights? Remember that this is a digital circuit, and only certain discrete values of calibration weight are allowed. In this case, we have to set a protocol that determines which of the values is most acceptable. The value of the last trial prior to the overweight trial is often selected for reasons of trade and the same is true of most SA ADC circuits.

Figure 18-10 shows the basic circuit of an SA ADC, while Fig. 18-11 shows a typical timing diagram for a hypothetical A/D conversion. In the SA type of ADC circuit, the scale is replaced with a specially constructed shift register, and the DAC represents the different values of weight applied to the "scale." The major components of the SA circuit are shift register, DAC, a set of output latches (one for each section of the shift register), and a control logic section. The output of the latches drive the inputs of the DAC.

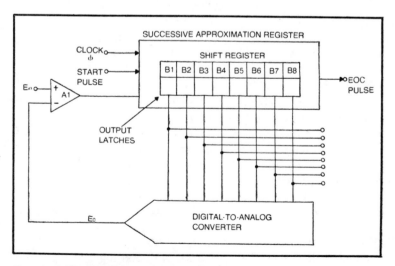

Fig. 18-10. Block diagram to the SA A/D converter.

A *start* pulse to the control logic section resets all latches to zero, and resets all shift register sections to zero. On the next clock pulse (T1 in Fig. 18-11), the output of the register is 10000000. This word places the output of the DAC at exactly one-half scale ($\frac{1}{2}$-E_{fs}). If the unknown input voltage is higher than this value, then the control logic section sets latch B1 HIGH. On the next clock pulse (T2), the next output register is set HIGH. The output of the DAC responds to the input word 11000000, and produces a $\frac{3}{4}$-E_{fs} voltage. If, on any trial, it is found that E_{in} is less than E_o, the latch that is active for that trial is set LOW. But if the input voltage is more than E_o, then the latch is set HIGH.

Figure 18-11 is timing diagram of a three-bit successive approximation ADC making a conversion. Let us say that the full scale input potential is 1.0V and the unknown input voltage is 0.625 volts.

t1—The start pulse is received. Register B1 goes HIGH. The output word is now 100 and E_o = 0.5 volts. E_o is less than E_{in}, so the output latch for B1 is set for HIGH (i.e. logical 1).

t2—At the next clock pulse, register B2 is made HIGH. The output word is 110, and the output voltage E_o is 0.75 volts. Since E_{in} is less than E_o, the B2 latch must be reset to LOW. The output word at the end of this trial is 100.

t3—At the third clock pulse, register B3 is set HIGH and the output word is 101. The output voltage from the DAC (E_o) is 0.625 volts, so E_{in} = E_o. The B3 latch is set HIGH, so the output word remains 101.

t4—Overflow occurs. This tells the control logic that the conversion is over, so issure an EOC pulse. In most cases, the overflow *is* the EOC, but in those that use a *status* output, the logic changes the level of the status bit.

The three-bit converter of our example required four clock pulses to complete the conversion. If a servo type had been used, the conversion would have been faster, but this will not happen once the bit length climbs to the five-bit, or higher, level.

At one time, the SA ADC was considered difficult to design and expensive to implement in any given piece of equipment. The problem was the amount of digital logic required to make the shift register, latches and the control section. But today, there are several integrated circuited successive approximation registers (SAR) on the market. A designer who wanted to start from "scratch," then does not need a lot of digital logic to make the circuit work.

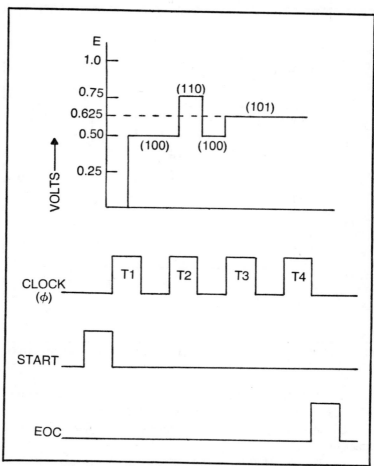

Fig. 18-11. Typical operating sequence.

Software ADC Techniques

Both servo and SA ADC circuits can be implemented in software, with very little additional circuitry required. The critical logic is performed in software routines in a computer.

Figures 18-12 and 18-13 show how computer I/O ports can be used to implement a software/hardware A/D converter. One complete output port is used to drive the inputs of an eight-bit DAC. A single bit is used to monitor the status of the comparator output. A program must be written that will generate a ramp at the output port. This is done by creating a loop that will increment a binary

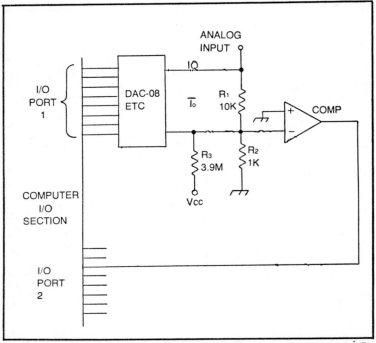

Fig. 18-12. Using a minimum of hardware, a software binary ramp (servo) A/D converter can be implemented.

word (usually held in a register) once for each pass through the loop. The elements required of the program are:

☐ Clear the accumulator (or other register used to hold the binary word).

☐ Increment the register by one.

☐ Output the binary word.

☐ Test the status bit at the input port

☐ If the comparator loop is HIGH, then the program loops back to (2).

☐ If the comparator output is LOW, then the program must break out of the loop and use the accumulator contents as the data word.

A flow chart of this procedure is shown in Fig. 18-13. This particular program example will implement the servo type of A/D converter.

An example of a successive approximation A/D converter using the combined hardware/software approach is shown in Fig.

18-14. This particular circuit uses the 8080A microprocessor by Intel, and a Precision Monolithics, Inc., DAC-08 digital-to-analog converter. A slightly different circuit, plus program modifications, however, would allow the use of any of the standard microprocessor chips. This particular circuit is a memory-mapped A/D converter, and is turned on by bit A15 of the microprocessor. This means that the address assigned to the A/D converter must be in the upper 32K of memory (convenient because many microcomputers do not have more than 32K of memory. A sample program that will operate the A/D converter is shown in Table 18-1, while a flow chart for a typical three-bit A/D conversion is shown in Fig. 18-15. Also see Fig. 18-16.

Software implemented A/D conversion schemes take CPU time, so are not used unless there is plenty of CPU time available. Note that some uP chips, and many different DACs, are now so cheap that it is almost cheaper to dedicate a microprocessor chip, small-scale ROM, and a DAC to make a "discrete" A/D converter.

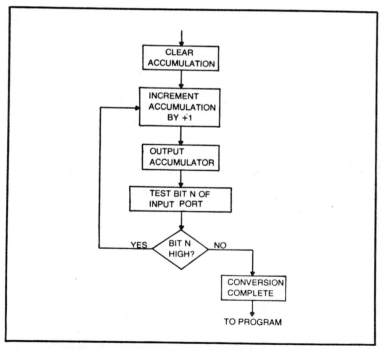

Fig. 18-13. Typical program flow chart for servo A/D converter in software.

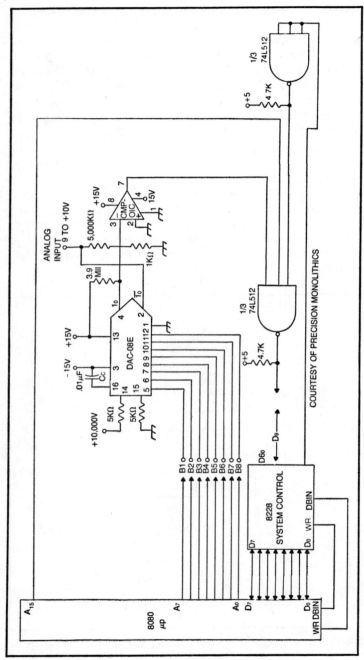

Fig. 18-14. Hardware for making an SA A/D converter in software with 8080 microprocessor.

Parallel (Flash) Converters

The parallel, or "flash," A/D converter is shown in Fig. 18-17. This converter is one of the fastest converter circuits known; its speed is limited only by the settling time of the voltage comparators used in the circuit. But this is not a free gift, because the output code is not one of the standard binary codes.

The flash converter circuit consists of $(2^n - 1)$ voltage comparators that are biased by a reference potential (E) through resistor voltage divider. Each comparator receives a bias that is 1-LSB different from the comparators above and below.

The very rapid conversion time is somewhat slowed by the need to decode the output prior to using it in an ordinary binary application. But the increased speed is so much that a few companies offer commercial flash converters in IC and hybrid forms.

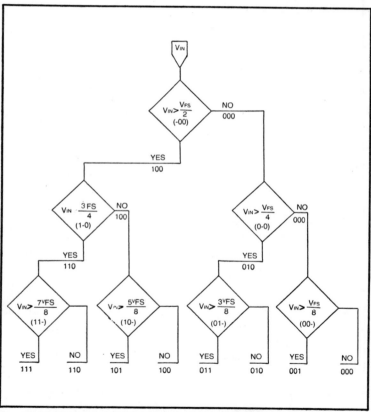

Fig. 18-15. Algorithm for making a three-bit SA A/D conversion.

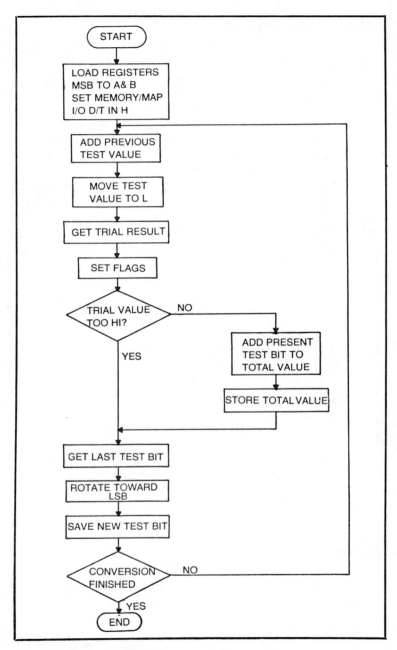

Fig. 18-16. Program flow chart.

Voltage-to-Frequency (V/F) Converters

A voltage-to-frequency converter is a type of A/D that converts the input voltage to a proportional frequency. It is, then, essentially a voltage-controlled-oscillator, or VCO (see Fig. 18-18).

There are two basic ways to use the VCO in A/D converter service. One is to use a computer program to measure the frequency of the output. The VCO signal is applied to one bit of the computer input port. A computer program can be written that will measure the *time* between successive input pulses, so that it can calculate the frequency; hence, the voltage. It is necessary only to know the scaling factor.

A second way to use the VFC is to gate its output into a binary counter. The gating pulse allows pulses from the VCO to input to

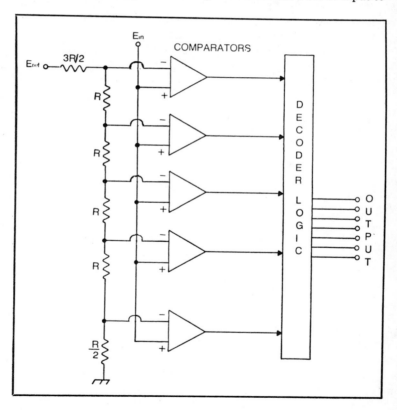

Fig. 18-17. Parallel, or flash, A/D converter.

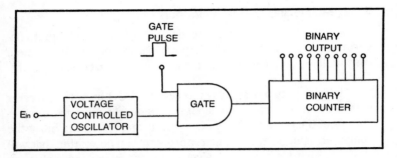

Fig. 18-18. V/F converter.

the counter, and the counter output then becomes the binary word proportional to the input voltage.

VFC A/D converters are used where serial data transmission is mandatory, or where simplicity is dictated. But the technique is limited by dynamic range (although some companies offer VFCs with 1000:1 frequency change over the input voltage change).

Chapter 19
DC Power Supplies For
Digital Equipment

Digital electronics equipment often has much tighter power supply specifications than some other types of circuit. Regulation of the voltage to within very tight limits is required. We might also find things like *overvoltage protection, overcurrent limiting* (shutdown), *thermal protection*, and something called *safe operating area protection*. The latter, incidentally, refers to the collector power dissapation capability of the regulator transistor. In this chapter, we will discuss the basic principles underlying low-voltage DC power supplies, some different types of power supply circuits commonly used in digital equipment, and some problems/troubleshooting notes.

BASIC PRINCIPLES OF DC POWER SUPPLIES

The basic DC power supply consists of a transformer, rectifier, filter, and regulator. Other features are added to the supply for some special purposes, but these sections comprise the basic core for all digital power suppliers. In general, power supplies for digital equipment are low voltage designs, and may produce current levels that are quite substantial.

Transformers

The transformer (Fig. 19-1) reduces the 115-volt AC power mains potential *down* to a level compatible with the low voltage requirements of the digital circuitry. Such transformers are called step-down transformers, and obey the following relationship:

$$E_p/E_s = I_s/I_p$$

Note from this equation that the current ratio is the inverse of the voltage ratio. A transformer that steps voltage down will show

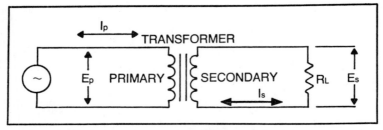

Fig. 19-1. Transformer with voltages and currents defined.

an apparently equal step-up of current. This concept can, unfortunately, become confusing to the new student who does not yet appreciate what is happening. In reality, secondary current I_s is determined only by secondary voltage E_s and the load resistance (R_L) connected across the secondary winding. The current step-up notion probably comes from the way this equation is often presented:

$$E_p I_p = E_s I_s$$

This equation tells us that I_p will vary as changing load conditions cause I_s to vary; as I_s goes up, I_p will also go up to keep the equation balanced.

Note that neither of the two forms this equation is presented have any loss terms. Real 60-Hz power and filament transformers are very efficient devices. True, there are losses involved, but most commercially available transformers are from 95 to 99.5 percent efficient, so the use of "lossless" equations is a justified simplification. The error terms are very small; so small as to be negligible.

Transformers are specified by their primary and secondary voltages, the secondary current, and the primary VA rating; i.e., the *volts times amperes* Note that VA is the same as watts *only* in purely resistive circuits. In reactive circuits, an additional *power factor* is present, and it is equal to the cosine of the phase angle between I and E.

In most cases, the primary voltage rating will be 110 volts AC, 220 volts AC, or will be selectable between 110 and 220. In the latter case, the transformer will have two primary windings. They are wired in series for 220V applications, and in parallel for 110V applications.

The secondary voltage will be determined by the turns ratio of the transformer. Also, the secondary must be center tapped for some rectifier systems, but not if a *bridge rectifier* is used.

All voltages specified on transformers are rms (root mean square). But note that both peak (E_p) and peak-to-peak ($E_{p\text{-}p}$) become important in the design of DC power supplies. The relationships between E_{rms}, E_p and $E_{p\text{-}p}$ for the sine wave are shown in Fig. 19-2.

The previous equations show that the primary VA and the secondary VA are equal at all times. But the primary VA rating is usually specified for the entire transformer. The VA rating is the maximum VA that can be obtained from the transformer without danger of overloading it. The reason for specifying the primary VA rating is that it is usually wound closest to the core, so cannot easily dissipate heat. As a result, the primary winding is usually easier to overheat. The primary VA rating must not be exceeded! While some manufacturers of equipment try to get away with it, this is not a smart practice. Failure to observe the primary VA rating may constitute a fire hazard and will almost certainly reduce the life of the equipment. Note that only transformers built to military specifications can normally be overloaded without danger.

Rectifiers

A rectifier (Fig. 19-3) is a device that will pass current only in one direction. This feature allows them to be used to convert bidirectional AC to unidirectional pulsating DC. Figure 19-3A shows a simple half-wave rectifier circuit. On the positive alterna-

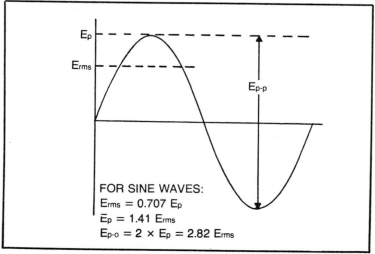

FOR SINE WAVES:
$E_{rms} = 0.707\ E_p$
$E_p = 1.41\ E_{rms}$
$E_{p\text{-}o} = 2 \times E_p = 2.82\ E_{rms}$

Fig. 19-2. Sine wave showing relationships.

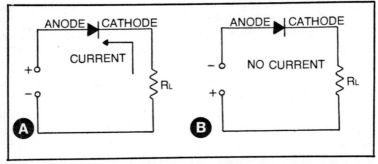

Fig. 19-3. Rectifier forward biased allows current to flow at A, and rectifier reverse biased and no current flows at B.

tion of the AC cycle, the diode rectifier anode is positive with respect to the cathode; the diode is forward biased, so current will flow. But on negative alternations, the anode is negative with respect to the the cathode, so the diode is reverse biased (Fig. 19-3B). No current will flow. The output waveform (Fig. 19-4) is unidirectional, but is not pure DC.; it is a form of a DC called pulsating DC. Because of the missing half cycles, the average voltage at the output of a half-wave rectifier is only 0.45 times the applied rms voltage ($E_{\frac{1}{2}} = 0.45E_{rms}$). The missing half cycles also makes it necessary to have a transformer with a VA rating that is 40 percent higher than the required voltage times current.

The half-wave rectifier, then, is terribly inefficient because it does not use the entire AC input cycle. A rectifier that *does* use the entire AC cycle is called a *full-wave rectifier*. Examples of the two common full-wave rectifier circuits are shown in Figs. 19-5 through 19-9.

Figure 19-5 shows a simple full-wave circuit using two rectifier diodes and a transformer that has a center-tapped secondary winding. The center-tap is designated as the zero reference point and is usually grounded. On any given half cycle, one end of the secondary will be positive with respect to the center tap, while the other end is negative with respect to the center tap. The situation reverses on every half cycle. We have designated the two ends of the secondary winding as points A and B for sake of the discussion. One one-half of the cycle, point A is positive and point B is negative. On the other half of the cycle, point A is negative and point B is positive.

Consider the situation in which point A is positive and point B is negative. In this case, diode D1 is forward biased, and diode D2

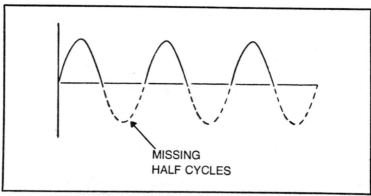

MISSING
HALF CYCLES

Fig. 19-4. Output wave from of half-wave rectifier.

is reverse biased. Current will flow from the center tap, through load resistor R_L, diode D1, and back to the transformer at point A.

On the second half of the AC waveform the situation is reversed; point A is negative and point B is positive. In this case, diode D1 is reverse biased and D2 is forward biased. Current will flow from the center tap, through the load resistor, diode D2 and back to the transformer at point B.

It is important to note that the current flows through the load resistor *in the same direction* on both halves of the AC cycle. This fact accounts for the double-humped waveform (Fig. 19-6) that is characteristic of full-wave rectifiers. The average output voltage from a full-wave rectifier is 0.90 times the applied rms voltage;

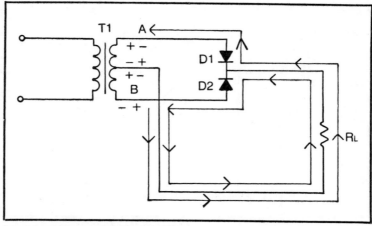

Fig. 19-5. Normal full-wave rectifier circuit.

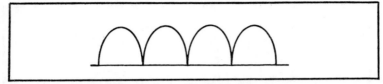

Fig. 19-6. Output waveform from a full-wave rectifier.

exactly twice the output that was obtained from the half-wave circuit.

Note that the more efficient full-wave circuit is universally preferred in digital (and most other) electronic equipment. Only in transformerless designs, such as might be seen frequently in consumer electronic products, is the half-wave circuit preferred.

A full-wave *bridge rectifier* is shown in Fig. 19-7. This circuit does not need the transformer center tap, but this is only at the expense of two additional rectifier diodes.

As in the previous case, the two ends of the transformer are designated A and B. On one half of the input AC cycle point, A is positive with respect to point B. In this case, diodes D1 and D2 are forward biased, while D3 and D4 are reverse biased. Current flows from point B, through diode D2, load resistor R_L, and diode D1. It returns to the transformer at point A. On the second half of the AC cycle, the situation is reversed. Point A is now negative with respect to point B, so D3 and D4 are forward biased and D1/D2 are

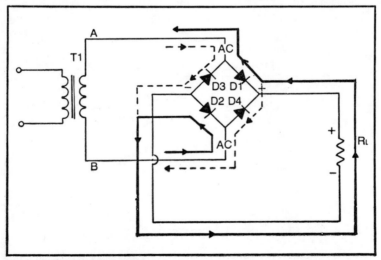

Fig. 19-7. Full-wave bridge rectifier.

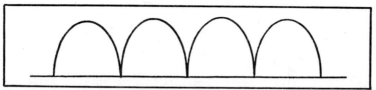

Fig. 19-8. Output waveform from a full-wave bridge rectifier.

reverse biased. The current flow is from point A, through diode D3, load R_L, diode D4 and back to the transformer at point B.

Again, the critical factor for using the full AC sine wave is that the current flows through the load resistor in the same directions for both halves of the alternation. The average output voltage is the same as for the other full-wave circuit; i.e., 0.9 times the applied rms voltage. See Fig. 19-8.

The transformer used with the bridge circuit need not be center-tapped. The zero potential reference point is designated as the junction of the anodes of D2 and D3. This point is labeled negative (−), while the junction of D1 and D4 is labeled positive (+). Some prepackaged bridge rectifiers (see Fig. 19-9) have the DC terminals labeled with the "+" and "−" symbols, and the AC terminals labeled with either the letters "AC" or the sinewave symbol as shown.

One advantage of the bridge rectifier is that it will produce an output voltage that is twice that of the simple full-wave rectifier. This is due to the fact that the entire secondary of the transformer is used on each half cycle, instead of just half of the winding. But this is not a free gift, because we must still guard against exceeding the primary VA rating. The secondary current rating of most center-tapped transformers is assigned with the assumption that

Fig. 19-9. Circuit symbol sometimes used for bridge rectifiers.

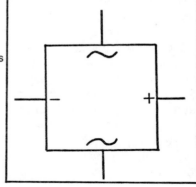

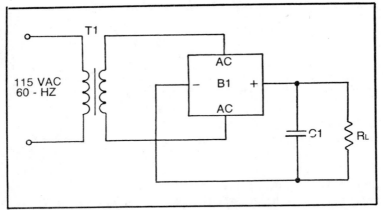

Fig. 19-10. Simple brute force capacitance filter.

the simple full-wave circuit is being used. If a bridge circuit is used, instead, then only one-half of the rated current is actually available without exceeding the VA rating. The full rated secondary current can be obtained only if the transformer is designed specifically for use with bridge rectifiers.

Filters

Most electronic circuits cannot use pulsating DC as derived derived directly from the rectifier. They require instead pure (or nearly pure) DC. The pulsations are called *ripple*. Half-wave rectifiers produce a ripple component of 120 percent, at a frequency equal to the line frequency (60 Hz in the US). Full-wave rectifiers, on the other hand, produce a 48 percent ripple component at a ripple frequency of twice the line frequency. In the US, this means 60 Hz × 2, or 120 Hz.

A *filter* circuit smooths the pulsations to produce nearly pure DC. In the simplest case of the capacitor filter (Fig. 19-10), a single capacitor C1 is connected across the rectifier output, in parallel with the load. Although more complex filter circuits are known, most digital instruments use the simple circuit shown.

The action of the filter capacitor is shown graphically in Fig. 19-11. The heavy lines indicate the pulsating DC waveform without the filter (C1 were disconnected), while the dotted lines show the output with the filter capacitor connected. Capacitor C1 will charge to approximately E_p, but after the peak has passed, the charge will return to the circuit. The effect of returning the charge from C1 to the circuit is to fill in the area between the pulses as

Fig. 19-11. Output waveform with a capacitance filter.

shaded the portion of the waveform. The filter reduces the ripple to a low percentage.

The value of C1 is not absolutely critical, only its *minimum* value is important. Higher values may be used with no deterioration of performance. In general, though, we require higher minimum values in half-wave circuits than in full-wave circuits. This is because of the lower ripple frequency in half-wave circuits. For most low voltage full-wave power supplies, the value of C1 will be 1000 μF/A or greater (some authorities prefer 2000 μF/A). In really low-current applications (100 mA), however, this rule is modified to "no less than 250 μF, with 500 μF preferred." A 10-A DC power supply for a small microcomputer, then, should have a 10,000 μF filter capacitor. Some authorities, of course, would say 20,000 μF.

The circuit in Fig. 19-10 is the simplest form of DC power supply. But note that it is almost incapable of totally eliminating the ripple factor. Even if a very high value filter capacitor were used (and take a lot of space), the power supply would show some ripple in the output voltage.

A voltage regulator circuit will reduce the ripple factor almost to zero, even though the regulator's main function is to maintain a constant output voltage. One power supply manufacturer used to sell 6 to 16 volt bench power supplies with a very low ripple factor.

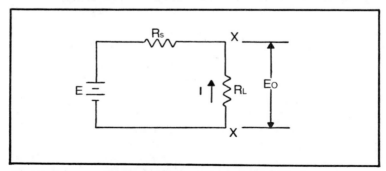

Fig. 19-12. Source resistance is a cause of regulation problems.

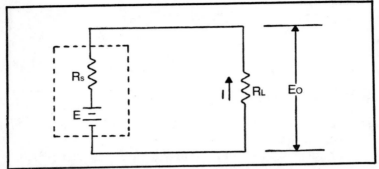

Fig. 19-13. Equivalent circuit of a power supply.

The salesmen would tell you that it obtained this spec because the filter capacitor (15,000 μF) was "amplified" by a special electronic circuit to 1 farad. But in reality, the ripple reduction was caused by a voltage regulator circuit that reduced the ripple as much as would be obtained from a 1 farad capacitor. There was no "capacitance amplification," involved; only a 10 KμF capacitor and a voltage regulator!

Voltage Regulators

All DC power supplies have a certain amount of internal source resistance, R_s. When a load current is drawn from the power supply, a voltage drop, IR_s, will occur across the source resistance.

The actual value of the source resistance can be determined from Ohm's law. The voltage used in this calculation is the open-terminal power supply voltage obtained by disconnecting load resistor R_L and then measuring the output voltage from the supply when $I = 0$. The current used in the calculation is the short circuit current; i. e. the current that flows when the output terminals are shorted together. *Don't actually try to make this measurement!* Most DC supplies cannot long survive a direct output short without extensive damage. An alternative method for measuring R_s is this:

$$R_s = \frac{E - E_o}{I}$$

where R_s is the source resistance in ohms, E is the open-terminal output voltage, E_o is the output voltage existing existing under load R_L, and I is the current drawn when E_o is measured ($I = E_o/R_L$).

Voltage regulation is a measure of how stable the output voltage remains between load and no-load conditions. Fig. 19-13 shows the mechanism that causes the output voltage change. Internal resistance R_s is effectively in series with load resistance R_L. The output voltage, then, can only be a fraction of the open-terminal voltage because R_s and R_L form a voltage divider. The output voltage at full load will be

$$E_o = \frac{ER_L}{R_s + R_L}$$

Voltage regulation is usually specified in terms of a percentage, which is calculated from the equation:

$$\%\text{Regulation} = \frac{(E - E_o)\,(100\%)}{E}$$

There are four basic forms of voltage regulator circuit: *Zener diode, Zener-referenced series-pass, feedback*, and *switching*. Others are often merely different forms of these basic types.

Zener diodes have a property that allows them to be used as voltage regulators. Fig. 19-14 shows the curve for a typical Zener diode. In the $+E$ region, the diode is forward biased and will behave exactly like an other PN junction diode. But in the reverse region ($-E$), the diode is reverse biased. The Zener diode is a little different from ordinary silicon diodes in this region. From zero volts, to a point marked V_z, the diode *is* like other diodes. No current, or only a tiny leakage current, flows. But at V_z the diode

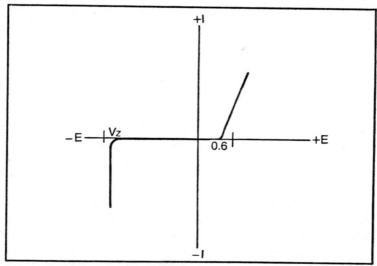

Fig. 19-14. Zener diode characteristic.

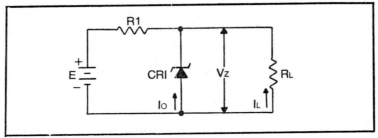

Fig. 19-15. Zener diode voltage regulator.

breaks down and conducts a reverse current. The value of V_z tends to remain constant for a wide range of applied voltage and reverse current level. It is to V_z that the Zener will regulate the output voltage.

Figure 19-15 shows a typical low-current regulator based on the Zener diode. Diode D1 is connected in parallel across the load. Resistor R1 used to limit the current to a level that is safe for the Zener diode. Without R1 the diode would burn out.

The simple Zener voltage regulator circuit is used only for light duty work. A general rule-of-thumb is that the load current must be held to 10 to 20 percent of the Zener current; hence, the limitation to low current applications.

A better solution is shown in Fig. 19-16. Here the Zener diode is used as a low-current reference voltage source in a series-pass circuit. In this circuit, transistor Q1 is the actual control device, and D1 is a reference. The output voltage is given approximately by:

$$E_o = V_z - V_{be}$$

The maximum output current is approximately the normal Zener load current (I_b in this case) multiplied by the beta of transistor Q1. This assumes, of course, that neither the maximum collector current, nor the maximum collector dissapation ratings are exceeded.

Another form of series pass voltage regulator is shown in Fig. 19-17. This circuit is called a feedback voltage regulator and depends upon comparing the output voltage E_o with a reference potential E_{ref}. In this circuit, the bias applied to series-pass transistor Q1 is determined in part by the output of unity gain differential amplifier A1. Amplifier A1 is often called an *error amplifier*. It is a differential amplifier because its output is proportional to $E_{ref} - E'_o$. If voltage E_o changes from the value set by E_{ref}, as when a load

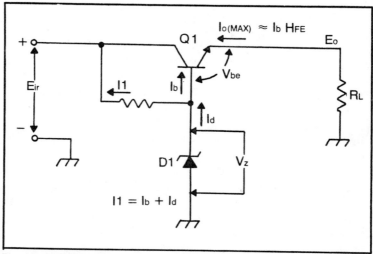

Fig. 19-16. Series-pass voltage regulator circuit.

current change occurs, then the sample E_o' applied to the error amplifier also changes. The amplifier output will respond by changing an amount and direction that tends to cancel the error.

A *switching regulator* is shown in simplified form in Fig. 19-18. In this circuit, a switching transistor (shown as an SPDT switch in Fig. 19-18) is used. The transistor must have a very low value of $V_{ce(sat)}$ for the circuit to operate properly. The switching regulator operates at a high frequency, such as 15 kHz. The LC network forms a low-pass filter to reduce the high-frequency component. Coil L must have a very high Q. If the transistor switch is turned on, then the full input voltage is applied to the output. But if the

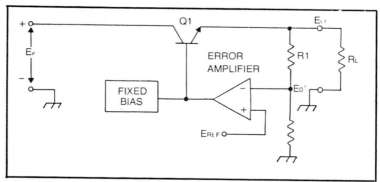

Fig. 19-17. Feedback amplifier voltage regulator.

transistor is turned off, then zero volts is applied to the output. The circuit works by pulsing the switch on and off for periods that reflect the output voltage. In some circuits, the duty cycle is adjusted so that a 50 percent on-time results in the proper output voltage. The on-time of the switch is determined by the output of a pulse width modulator, which is in turn controlled by the output of the error amplifier. If the output voltage drops, then the PWM increases the on time of the switch, causing more energy to flow into the low pass filter. But if the output voltage rises, then the on-time of the switch decreases.

In general, switching regulators (of which, Fig. 19-18 is but one type) are considerably more efficient than series pass circuits. Keep in mind that the switching regulator may, at first glance, look much like a series-pass circuit because Q1 is in series with the V_{in} supply. But the tell-tale sign is the use of a low-pass LC filter at the output and a pulse-width modulator.

Actual IC and Hybrid Regulator Devices

The easiest way to obtain low-voltage DC power supplies of 1 ampere (and possibly to 5 or 10 amperes!) is to use a fixed output voltage three-terminal IC regulator. These devices, which are housed in transistor packages, can be obtained in standard voltage level values up to 24 VDC.

Three-terminal IC voltage regulators are available in several different, but similar, families. Three general catagories are available: 100 mA, 750-1000 mA, and 1000-1500 mA. There are also several individual types available in current ranges up to 10 amperes. Figure 19-19A shows several of the common case styles used for voltage regualtors. Note that they are all transistor cases. The letters, which are used in the type number designation for the regulator, tell us the case style and the approximate current limitations. Consider the LM-309 device for example. There are at least two different LM-309s on the market: LM-309H and LM-309K. Both are +5 VDC regulators. The LM-309H, which is housed in a TO-5 transistor package, is a 100 mA device. The LM-309K, on the other hand, is in a TO-3 package so can dissipate more power. The rating of the LM-309K is 1 ampere in free air. Up to 1.5 amperes (some manufacturers claim 2A) can be drawn if the device is properly heat-sinked. The T-package, incidentally, is the transistor TO-220 plastic package (sometimes called P-66), and will dissipate up to 750 mA in free air and 1 ampere if properly heat-sinked.

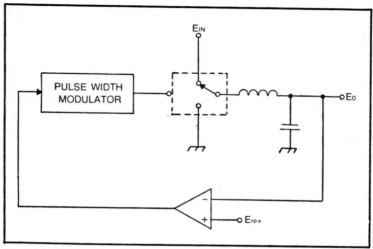

Fig. 19-18. Switching regulator.

The most common three-terminal IC devices are the LM-340 and 78xx series regulators. For LM340 devices, the voltage rating is given by a suffix separated from the main number by a hyphen, while the current rating is given by the case-style letter:

LM-340 (x) − (yy)

The case-style letter is placed where (x) is shown, while the voltage rating is placed where we show (yy). Only the letters H, K or T are used. The LM-340K-12, then, is a 12V positive voltage regulator in a TO-3 (i.e. K-style) package. Because of the package style, it is capable of passing up to 1 ampere. The LM-340T-15, on the other hand, is a 15V positive regulator in a T-style (750 mA) package.

The 78xx series replaces the "xx" with the voltage rating. A 7805, then, is a 5 volt positive regulator, while the 7812 is a 12V positive regulator. It is not usual to include the case style as part of the type number in this series of devices. Note, also, that Motorola oftentimes uses their own case style (the one with the hole in the middle) for the plastic package devices. These may, however, bear the 78xx designation.

Negative regulators are also available in the same package styles and voltage ratings. These are given the designations 79xx and LM-320. The same protocols are followed for the type number. An LM-320K-15, for example, is a 15-volt, negative voltage regulator in a TO-3 (1 ampere) package. Note that there is a pinout difference between positive and negative voltage regulators! They are *not* pin-for-pin compatible. If you try to connect a negative voltage regulator into a circuit where the positive regulator had been used before, it is simply not enough to reverse the polarity of the input in order to obtain the required negative voltage output. You must also swap the correct pins on the IC regulator.

The basic circuit for positive voltage regulators is shown in Fig. 19-19B. Transformer T1, diodes D1 and D2, and filter capacitor C1 are the same as in any unregulated DC power supply. Note also, that a bridge rectifier may be substituted for D1/D2; in fact, it is in most cases. Capacitor C1 is the filter capacitor, and should not have a value less than 2000 μF/A. The function of capacitors C2 and C3 is to improve the immunity of the circuit to noise impulses. Capacitor C4 is optional, but is used to improve the transient response of the output circuit. The value of C4 should be approximately 100 μF/ampere.

Until relatively recent times there were no voltage regulators on the market that were capable of delivering more than 1 ampere (without loads of heatsinking!). But now, there are several on the market. One of the first was the Fairchild Semiconductor LM-323 (TO-3 package) that was capable of delivering 3 amperes. Fairchild and Motorola now offer several models in the 5 ampere range, and at least one type is available that will produce 10 amperes. Lambda Electronics (515 Broad Hollow Road, Melville, NY 11746) produces a series of three-terminal IC regulators in the TO-3 package. They offer 1.5 ampere, 3 ampere, 5 ampere, and 8 ampere models. These are the LAS-15xx, LAS-14xx, and LAS-19xx series, respectively. As in the 78xx series, the voltage number replaces "xx" in the type number. An LAS-1905, then, is a 5 volt regulator able to deliver up to 5 amperes.

An exception to the Lambda numbering system is the LAS-19CB, which delivers 13.8 VDC at 5 amperes. This device is designed mostly for "12 volt" DC power supplies for operating CB rigs from the AC power mains. Note that most "12 volt" automobile electrical systems are actually 13.8 to 14.4 volt systems.

All three-terminal regulators have a certain minimum and maximum input-output differential voltage ($E_{in} - E_o$) rating. A typical number is 2.5 volts, although some ask for as much as 3 volts, or as little as 2 volts differential. The maximum voltage is related to the amount of power that can be dissipated safely in the collector of the internal series-pass transistor. For most devices, the maximum input voltage will be in the 30V to 40V range. The

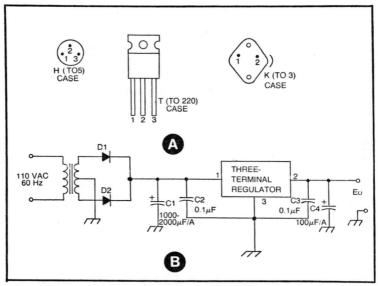

Fig. 19-19. Three-terminal regulator case styles are similar to transistor cases at A, and typical circuit for positive regulators at B.

minimum I/O differential is the smallest differential that will allow the regulator to operate properly, while the maximum voltage is the level that may cause damage to the device if exceeded. Note that some hybrid regulators that deliver very high currents (i.e. more than 15 amperes) will not deliver full output current if the voltage is operated near the maximum. These devices will have a dissipation rating in watts that must not be exceeded. We must, therefore, derate the current specification if the device is operated near the maximum allowed input voltage. Consider a 5-volt, 5-ampere regulator, for example. It may have a 50-watt dissipation rating, and a 35-volt input voltage. The power actually dissipated is:

$$P_d = I(E_{in} - E_o)$$

$$= I(E_{in} - 5)$$

So, if the maximum input voltage is the allowed 35 volts,

$$P_d = (5A)(35 - 5)$$
$$= (5A)(30) = 150 \text{ watts!}$$

From the above it is clear that we cannot expect to draw the full rated output current if we operate the device near the full allowed

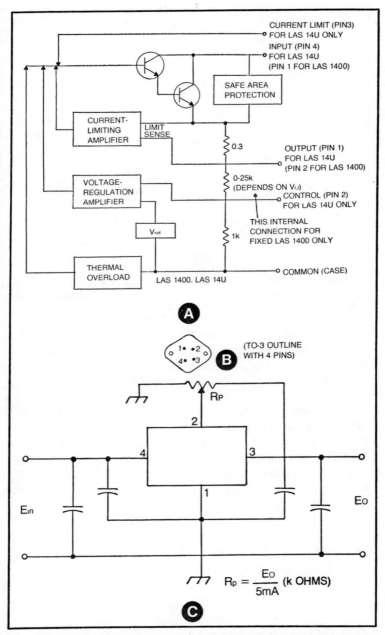

Fig. 19-20. Lambda four-terminal regulator internal circuit block diagram at A, and case pinouts at B, and a typical circuit at C.

input voltage. We can find the maximum allowed output current under those conditions by rearranging the equation to solve for I when E_{in} is 35 volts (or whatever voltage is actually used) and P_d is 50 watts. Similarly, we can also find the maximum allowed input voltage that will permit us to draw the full 5 amperes by rearranging the equation to solve for E_{in} when I is 5 amperes and P_d is 50 watts. In the case shown, the full rated output current is available only for input voltages less than 15VDC. The same reasoning also applies to other types of voltage regulator.

Lambda Electronics also offers a line of four-terminal IC regulators. These are similar to three-terminal devices, but have one additional terminal that permits us to adjust the output voltage to any point within a given range. An example of this circuit is shown in Fig. 19-20. The device is housed in a four terminal version of the TO-3 transistor case. A simplified internal circuit diagram is shown in Fig. 19-20A. Note that it is a feedback type of regulator, and has thermal, safe operating area, and overload protection. The latter becomes very important to the survival of the power supply if a catastrophic short circuit develops. The series-pass element inside of this regulator is a Darlington pair of transistors. This allows us to take advantage of the extremely high beta possible in the Darlington configuration ($H_{fe(Q1)} \times H_{fe(Q2)}$).

The case style for this regulator and the pinouts are shown in Fig. 19-20B. Since the case is similar to TO-3, we find that TO-3 heatsinks will work nicely with the regulator, provided that the extra holes are drilled to accommodate the other two terminals.

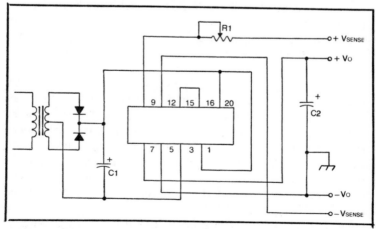

Fig. 19-21. Circuit for a Lambda hybrid regulator circuit.

The Lambda four-terminal regulators can be set to any voltage that is within their range using potentiometer R_p (see Fig. 19-20C). The range for most models of this series is about 4 to 30 volts. The LAS-15U, LAS-14U, and LAS-19U produces output currents of 1.5, 3 and 5 amperes, respectively. With the exception of the control pin and the potentiometer, the four-terminal regulator uses the same general circuit as the three-terminal devices.

High-current power supplies of over 5 to 10 amperes are a little tricky to design properly. You can't expect to just slap a Zener diode into the base circuit of a high gain transistor and make it work properly. We suddenly run into thermal problems and certain other difficulties. Additionally, it is too easy to buy ready-built, hybrid, voltage regulators in current levels up to around 30 amperes. The Lambda models in this range are:

LAS2xxx	5 amperes	85 watts
LAS3xxx	10 amperes	140 watts
LAS4xxx	10 amperes	170 watts
LAS5xxx	20 amperes	270 watts
LAS7xxx	30 amperes	400 watts

The LAS5205 is a 20 amperes, 5-volt regulator that will dissipate over 200 watts. Similarly, the LAS7215 produces 15 volts at almost 30 amperes.

Figure 19-21 shows an example of a circuit using a Lambda hybrid voltage regulator. The value of the voltage adjustment trimmer potentiometer should be *at least* ($0.25E_o \times 1000$ ohms/volt).

The exceptions to this rule are the 5-volt models, in which the minimum value of the resistor is 3000 ohms. In actual practice, the minimum value for any given specimen of voltage regulator may be a lot higher. In a power supply which I built for my microcomputer, the acutal value required was in the 11,000 - 14,000 ohm range.

Note that both pins 1 and 20 are positive-input-voltage terminals. This can lead you astray in some cases because they have different input-output voltage requirements. Pin No. 1 will operate with a +2.5 volt I-O differential, while pin No. 20 requires a +7.5 minimum differential! Pin No. 1 is the normal high-current input, while pin No. 20 powers the internal control circuits. If a device uses a common +8-volt, unregulated high-current power source for the regulator, then another low current supply must be designed and built for pin No. 20. Fortunately, the current requirements here are low, and most microcomputers have a +12 to +15 VDC supply which can be used for pin No. 20.

OVERVOLTAGE PROTECTION

Most integrated circuits must operate within certain voltage limits. While the CMOS devices have a relatively wide operating voltage range, the TTL device is very limited. In no instance must the voltage rise above 5.6 volts DC, or damage to the device will result. And it becomes damaged in a relatively short period of time! Most digital power supplies have +8 volts on the input side of the regulator, and it is a high current source. If the series pass element in the regulator shorts c-e, then this +8 volts is applied to the +5-volt line and will probably damage most of the TTL devices in the circuit. To prevent this catastrophe, use an overvoltage protection circuit at the output of the regulator. Although Lambda makes a two-terminal module for this purpose, most will use a discrete circuit such as shown in Fig. 19-22. This circuit is called an *SCR crowbar* because it is a brute force approach to solving the problem. High-current SCR D2 is connected in parallel across the output of the supply. It is normally turned off. Diode D1 is a +5.6-volt Zener diode. If the line voltage increases above +5.6 volts, then D1 conducts and thereby turns on the gate of the SCR. This action causes the SCR to turn on and short the output of the power supply. In most cases, there will be a series fuse or circuit breaker. In other cases, the primary fuse protecting the transformer will blow or the SCR will just simply sit there and absorb the excess current flow.

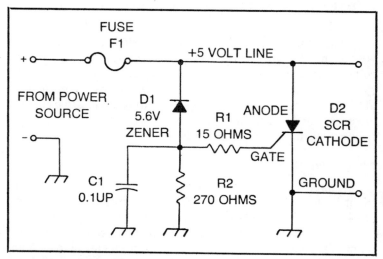

Fig. 19-22. SCR crowbar overvoltage protection circuit.

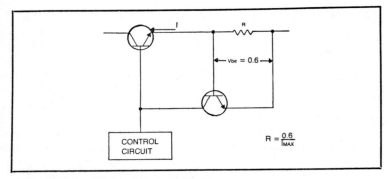

Fig. 19-23. Output current limiting circuit.

OUTPUT CURRENT LIMITING

A short circuit in the digital device can cause immense damage due to the high current nature of the power supply. A short on one of the PC boards in an unprotected piece of equipment can burn up all of the printed circuit lands in the +5-volt and ground lines all the way back to the power supply. In addition to that, it can destroy the power supply in quick order. The answer is output current limiting. A simple example of an output current limited circuit is shown in Fig. 19-23. Transistor Q1 is the regular series-pass element for the voltage regulator. Transistor Q2, however, is used as a shutdown switch for the current limiter. The bias for Q2 is obtained across a small series resistor that has a value of

$$R = 0.6/I_{max}$$

When the maximum current is exceeded, the base-emitter voltage needed to turn on Q2 is exceeded. This causes the collector-emitter path of Q2 to drop in resistance, effectively shorting together the base and emitter of Q1. A small increase over maximum output current causes "soft limiting," but as the current increases even more the "hardness" of the limiting increases rapidly until total shut off is reached.

TYPICAL POWER-SUPPLY CIRCUITS

Figures 19-24 and 19-25 show two typical power supplies that might be found in digital equipment. The circuit shown in Fig. 19-24 is a ±12-volt, low-current (1-ampere) power supply. It is based on the LM-340 positive and LM-320 negative voltage regulator ICs. Note that both of these circuits are identical to circuits presented earlier in this chapter, except for the arrangement of the transformer secondary and rectifier. At first glance, you see a bridge rectifier, but note that the transformer is center-tapped! This is not a contradiction, because one-half of the secondary and one-half of the bridge rectifier are used for each side of the power supply. The bridge is not full-wave, but is called a two-diode bridge, or half-wave bridge, rectifier.

The circuit in Fig. 19-25 is a 10-amperes, +5-volt DC source for the TTL circuitry in a mircocomputer. The rectifier (B1) is a 25-ampere type built in modulator form. The filter capacitor (C1) has a rating of 18,000 μF, but in many cases expect to find values up to 100,000 μF. In a version of this circuit that I built for my microcomputer an 80,000 μF unit was used. Transistor Q1 is the series-pass element, while Q2 is the current limiter. The heart of the circuit is a Motorola IC voltage regulator. This circuit device can be used alone at currents up to 500 mA, but needs a series-pass element at higher output currents. It is housed in a 10-pin transistor-like package that has physical dimensions similar to the small TO-66 power transistor case.

Note the *remote sense* terminals in this supply. This is very common on the +5-volt supply of digital equipment. It seems that, at the current levels expected, ordinary ohmic losses in the conductors between the output of the regular and the actual circuit boards can be quite high. They may look small on first glance. But when you consider the fact that many TTL devices begin to act up if the operating voltage drops below +4.75 volts, then it becomes apparent that millivolts are important. The sense line is to the error amplifier in the internal feedback regulator (inside the Motorola IC). The sense line is connected to the point on the PC board where we want to insure a +5-volt operating potential. In actual practice, this line is placed either at the farthest extreme, near the most critical board, or about ¾ the way between the supply and the farthest PC board.

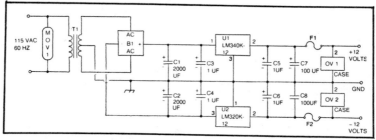

Fig. 19-24. Dual-polarity power supply.

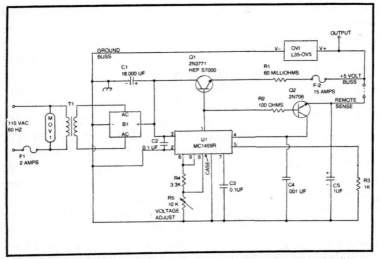

Fig. 19-25. High-output current, low-voltage power supply with sense line.

TYPICAL POWER-SUPPLY PROBLEMS

One service shop that kept records of the actual faults found in warranty equipment discovered that more than 60 percent of the total number of service jobs were either in the power supply itself, or generated symptoms that affected the power supply voltages. While this information will not surprise old timers in the electronic service business, it is good advice to the newcomer: Look at the power supply voltages *first*. Regardless of the actual numbers and percentages on any specific equipment, it is true that you can solve a lot of problems quickly by taking a brief look at the DC power supply, especially when symptoms appear weird.

Chapter 20
Test Equipment
For Digital Servicing

One of the first questions asked by anyone responsible for an electronic servicing business is: "What kind of test equipment is needed, and how much does it cost?" When color television first came on the scene, we learned about things like color-bar generators, vector scopes, and wideband oscilloscopes. But even then, the standard of the industry was recurrent sweep models for servicers. Triggered sweep oscilloscopes were regarded as too expensive, exotic, complicated, and all of above. But today, the typical TV servicer will use a recently designed triggered sweep oscilloscope routinely.

If you are already in the electronic servicing business, then you may own all of the test equipment that you need for some types of digital servicing. Indeed, only a small additional investment need be made, and that in small items.

The ordinary test equipment common to all electronic servicing will usually fill the bill in simple digital servicing. You will, for example, need a good portable voltmeter or multimeter. Most digital servicers prefer the digital multimeter types of instruments. If you have an "old fashioned" analog meter, however, don't rush out and buy a brand new digital job unless you want to, or need an additional meter. That old VOM/VTVTM will work just as well for the rough measurements needed in digital work. Keep in mind, though, that digital servicing tends to be field servicing. Some of the new digital multimeters are more suited to field work than most analog meters. They are thinner and cost less in battery power. Those that use CMOS ICs and a liquid crystal display will last a long time on a battery change—and will be easier to maintain.

If you have any bench power supplies, then they will be useful in digital electronics servicing providing that they are regulated.

An old car radio "battery eliminator" probably will not do the trick, however. Bench supplies that do not meet the specs can be upgraded by addition of a couple three-terminal voltage regulators (see Chapter 19). It is especially necessary to provide one or more sources of +5 VDC at currents up to 5 amperes. One service shop used a Lambda LAS-1905 (5 VDC at 5 ampere) three-terminal regulator inside of an old Delco 12-volt battery eliminator. A new heatsink on the back panel and a pair of heavy duty five-way (banana) binding posts did the trick. The input to the LAS-1905 was merely the main power from the Delco P1200, set to +8 volts (or more).

People entering the digital electronics service business cold might wish to contact the test instrument manufacturers for a package deal on needed instruments. Tektronix, Inc. (POB 500, Beaverton, Ore., 97005) offers their TM-500 line of low-cost modular test instruments (Fig. 20-1). The TM-500 instruments are designed for field service, as well as bench service, work. They are plug-in modules and can be configured in any of several mainframes. The example shown in Fig. 20-1 is the scopemobile version. The racks contain the power supplies and needed interconnections. Other racks include portable and luggage packages, presumedly for field engineers who travel a lot by airliner.

If there ever occurs a situation where I am told that only one type of test equipment or instrument will be allowed for me to perform a troubleshooting job, then my answer will be the *oscilloscope*. This is the one instrument that is truly versatile and almost indispensible. Color TV servicers long ago learned that they needed to update their old-fashioned "audio" range oscilloscopes that had served so well in B&W receiver troubleshooting. They bought 5 and even 10 MHz models for color service. In fact, those test equipment makers that specialize in equipment for TV servicers were prone to call these "wideband" scopes *Color TV oscilloscopes*.

In digital service work, the specs on that old 'scope will probably take another leap upwards. The vertical sensitivity is not too awfully critical, but that bandwidth may be important. The 4.5 MHz *color TV* scope might work out for much service work if clock speeds in the digital circuit are low and rise time of the scope is not critical. The rise time is related to, but not identical with, vertical frequency response. In general, on a properly designed oscilloscope, the rise time will be found approximately from

$$T_{(sec)} = 0.35/F_{(hz)}$$

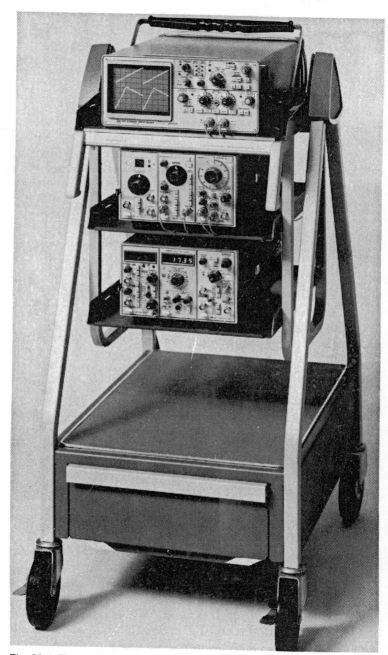

Fig. 20-1. Typical field service instrument package on scopemobile.

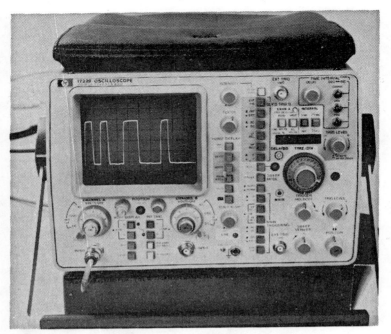

Fig. 20-2. Typical oscilloscope for digital servicing.

where T is the rise time of the vertical amplifier in seconds, and F
is the vertical amplifier frequency response in hertz.

As the frequency of the clock or the rise time of the pulses in a
digital circuit become higher, the requirements placed on the
oscilloscope used by the servicer becomes more critical. If you do
not own a good oscilloscope, then plan to buy one. Both Tektronix
and Hewlett-Packard will be glad to send you their catalog.

While some cheaper versions of scopes seem to have similar
specs, they simply do not perform as well as the ones that cost
more. A good scope is a tool of the trade.

Figures 20-2 through 20-4 show several different high quality
oscilloscopes used frequently in digital service work. Not all features of all models will be needed in any given shop, so one should
examine the catalogues, and request that a sales person bring one
out to the shop for an "audition" under field conditions. Some will
even leave a demonstration unit with you for a week or so, if you
seem like a good prospect.

Keep in mind the need for portability in your oscilloscope. If
most of the work will be permanent on-site assignments, then a
largish mainframe unit could be justified. But if all of the work will

be temporary service call assignments, please consider the poor slob who has to carry that darn oscilloscope four city blocks from the nearest parking space to the site! An extra heavy oscilloscope may tend to be left in the car more than not, with resulting backtracking lost time when the servicer finds the scope *is* needed.

LOGIC PROBES

After making an impassioned plea for you to rush right out to buy a multi-kilobuck oscilloscope, let me tell you a little story. One hospital where I worked had a complicated pulmonary test instrument that was made of mostly digital circuits. This instrument was a massive collection of digital circuits in a six-foot-high 19-inch rack panel. The field engineer *owned* a high quality Tektronix portable oscilloscope. But in two years, I never saw the scope anywhere but in the trunk of his car! When he came to service the equipment, he brought a briefcase tool box and two test instruments: a digital multimeter and a *logic probe.*

You ask just *what* the blue blazes is a logic probe? That's a fair question. It is a pencil-like probe (actually most look like a fat scope probe) that detects logic levels and clock pulses. Some will allow you to generate a pulse or a logic level, while others merely

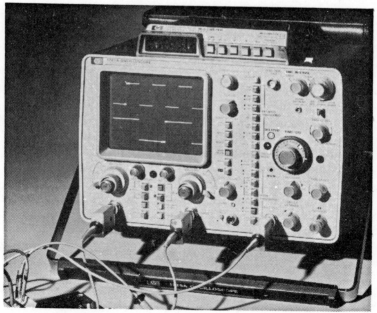

Fig. 20-3. Digital servicing oscilloscope.

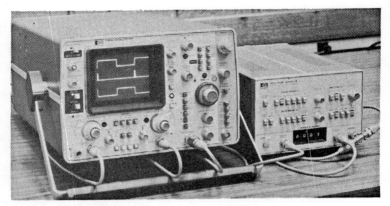

Fig. 20-4. Hewlett-Packard model 1741A oscilloscope.

detect and indicate activity in the circuit. Some will have three lights as indicators: HIGH, LOW and clocking. The clock detector is capacitor coupled so mere levels will not fire the lamp.

The reason why the field engineer for the pulmonary test equipment never needed a scope is that most of the circuit actions, except the clock, took place at slow speeds, or were merely levels that would come up or down as switches were manually closed. Most of the time, only the *existence* of a level or pulse needed to be confirmed. For that special case, a logic probe was superior to an oscilloscope. Such probes cost $38 to $295 and represent a low investment for a large return.

LOGIC ANALYZERS

In digital servicing, *timing* is often the keyword in finding problem. This means, *at least*, a dual-trace oscilloscope in most cases. Recently, though, the advent of multi-bit data paths, microcomputers, etc., have made it mandatory that we be able to examine the states of several bits *simultaneously*. If the events are periodic—repetitive, then an oscilloscope with a multi-channel electronically switched input circuit might be useful. But it is often the case that some event will trigger the circuit, and the events of interest to the servicer will be temporary. They will look like mere transients to the oscilloscope. A storage oscilloscope could trap some types, but they are very expensive as frequency increases. Another answer is to use a logic analyzer, as shown in Figs. 20-5 and 20-6. This instrument will capture and display on a CRT the states of the logic bits under examination. Then the servicer can analyze almost at leisure events that took place in nanoseconds.

This type of instrument is used in design labs and those service jobs involving complex instruments.

Some logic analyzers will display the levels (as shown in Fig. 20-5) in the fashion of a waveform timing diagram. But in computer work it is often the case that we want to see the binary words that passed along a data or address bus during the event. Some instruments will display as characters on a CRT screen, the binary words that were received. Some will allow a single bit to trigger the analyzer, and will then display, in either hexadecimal or binary form, the 256 words prior to and after the word existing when the trigger was received. The trigger word will be accentuated by excessive brightening of its representation on the CRT.

Another form of logic analysis involves signature patterns (Fig. 20-6). This technique is similar to vectorscope troubleshooting in color TV. But in the case of signature analysis, the instrument displays a pattern that indicates the performance properties of the circuit in response to a standardization stimulus—a certain pattern of operation within the circuit, such as a standard diagnostic program run by a computer being serviced. This technique becomes important when dealing with large-scale digital circuits, where it is difficult to analyze in the old stage-by-stage manner that worked so well in AM broadcast receivers of three decades ago.

Figure 20-7 shows a combination pulse generator and word generator system by Hewlett-Packard. This instrument is used in servicing and design of computers, where it will generate a given pattern of binary words. These can then be used to test the system.

Most logic analyzers are expensive instruments. For the majority of digital service jobs, they will prove unnecessary. But

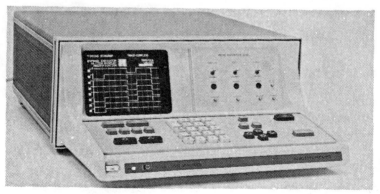

Fig. 20-5. Logic analyzer.

given the cost, it would be wise to weigh this factor carefully when taking on jobs that will require the instruments. The question is, as always, would it be cheaper given the incidence of service to buy a logic analyzer, or to be a board swapper and let the factory buy the logic analyzer (*not* an unaccepted practice in field service)?

TOOLS

If you intend to go into field service of classy digital equipment, then learn who Jensen Tools and certain of their competitors are. They are a sales organization that makes up field service tool kits in brief case tool boxes. This turns out to be both practical and a very necessary image builder. The tools themselves are pretty

Fig. 20-6. Hewlett-Packard logic analyzer.

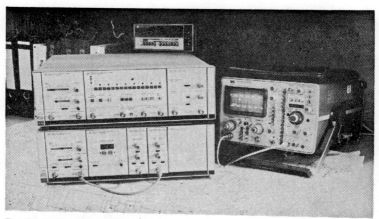

Fig. 20-7. Word and pulse generator.

much the same sort of thing that will be carried by a TV servicer in the tool/tube/module caddy. But image is very important when you are charging a lot for going out and fixing the equipment. How well you are regarded depends upon how the customer perceives you—and that affects the willingness to pay the freight! If you show up in dungarees and carrying a tool caddy, you are then a workman to be disdained. But is you show up in a suit, carrying the same darn tools in a tool brief case, you are seen as a field engineer worthy of your hire.

Besides, I have found that those brief cases are a lot easier to lug around than a bulky caddy, and their size limits the amount of trash that you carry into the customer's site. It is often too easy to go back for special tools than to carry everything everywhere.

Chapter 21
Some Common Problems

There are two principal areas where digital equipment can have recurrent problems: power supply and temperature. These two considerations cause many different types of problem. Many of the problems are either recurrent, toughdogs, or both.

Any piece of digital equipment that contains substantial amounts of TTL devices will generate a lot of heat. The high current requirements of *any* fast (high-frequency) logic family will mean a large amount of heat energy being dissipated. Unfortunately, many devices, especially when used in combination with (or synchronized to) other circuits, become flakey when the operating temperature goes above a certain limit. As a consequence, one may pay close attention to overheating problems. One stand-alone minicomputer, for instance, uses forced air blowers to keep the temperature down. The manufacturer has placed a hole in the cabinet to be used as a temperature test point. The service technician inserts a regular clinical medical fever thermometer into the hole, and allows it to come to equilibrium. The machine is considered to be operating within the temperature specification if the temperature at the test point does not exceed 102°F, or 38.9 °C).

Power supplies frequently encountered in digital equipment, especially larger systems, are usually equipped with thermal shut-down ciruits. These circuits will turn the power supply off when the over-temperature conditions exist. Also, series-pass and switching transistors tend to short out *collector-to-emitter* at temperatures over some point in the 125 to 200 °C range.
They are derated at temperatures above 30° C.

Temperature, therefore, can easily be a critical parameter in digital equipment. One might suppose that manufacturers would see to these matters in the original design of the beast. And usually they do! But, sometimes, Murphy's law takes over and some machine is designed with a deficiency in the design of the thermal characteristics; in other words, there was a screwup somewhere back in the engineering department. We also find instances where a customer will operate the equipment greater than originally specified by the manufacturer. Large, and even medium size, computers, for instance, are usually housed in a well-air-conditioned room. Some of those rooms get so cold that the operators must wear sweaters everywhere but in the immediate vicinity of the machine. If an instrument, machine, or computer is designed for a, say, 75 °F room, and is actually operated at a much higher temperature (say 90 °F), then the heat transfer away from the sensitive circuits is reduced. This will cause overheating and possible malfunctions.

The solution is to either reduce the ambient temperatures markedly, or increase the heat flow away from the circuits. In some cases, both strategies are needed. In some cases, the customer

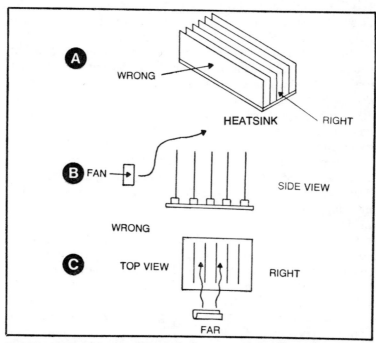

Fig. 21-1. Correct and incorrect airflow patterns across heatsink at A, incorrect airflow through stack of PC boards at B, and correct airflow at C.

might be willing to air condition the room. But if this is too expensive, impossible, or impolitic, then the other alternative must be examined.

In most cases, electronic equipment must be cooled with forced air. Ventilation holes must be placed so that air blown across the affected circuitry can escape and carry the heat with it.

There are right and wrong ways to eliminate heat from equipment. Fig. 21-1 shows several possible situations. In Fig. 21-1A a finned heat-sink, as might be found dissipating the heat generated by the series-pass element of the DC power supply (itself a major producer of heat) sensitivity function. The heat-sink should be physically placed so that air from the forced air blower travels between the fins. If it travels perpendicular to the fins, then it will not see maximum exposed surface area. Blowing air across the maximum surface area tends to maximize the heat transfer.

A similar situation exists with respect to printed circuit boards inside of a cabinet. The fan or blower should be positioned to blow across the component side of the printed circuit boards. Fig. 21-1B shows the incorrect way to position the fan, while Fig. 21-1C shows the correct placement. Note than in Fig. 21-1C, the fan should be placed to blow evenly across all of the boards. If one of the boards is known to run hotter than the others, however, then position the fan to blow predominently over that board. It might also be required to add an *additional* fan or blower to cover that particular board. There are also times when it is advisable to *increase the rating* of the existing fan. In one case, I replaced a 56 cu. ft./min. (7-watt) whisper fan with a 115 cu. ft./min. (14-watt) fan of the same series. This cured a difficult thermal problem for my employer.

Keep in mind that *heat rises*. Therefore, fans and blowers should be placed in the lower parts of the cabinet, and vent holes located near the top. Figures 21-2 and 21-3 show several situations involving equipment cabinets. In Fig. 21-2A, you see a small cabinet situation, such as might be found in desk-top, or bench top, instruments. The fan is mounted on the rear panel, and vent holes are placed along the top edge of the cabinet. Forced air circulates from the fan, upward through the printed circuit boards, and out of the cabinet via vent holes. Forty to 60 cu. ft./min is *usually* sufficient in small equipment cabinets, unless some element is present that would add to the heating normally associated with TTL devices.

Figure 21-2B shows a similar situation in which the fan is bottom mounted. Air forced up through vents at the top of the

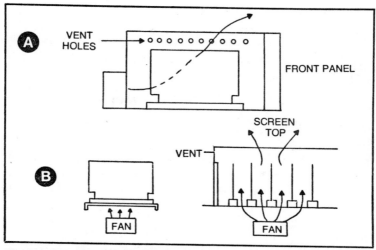

Fig. 21-2. Correct airflow through PC card cage when side mounted at A, and bottom mounted fan at B.

cabinet. Some enclosures have a vent port, or series of holes, along the top edge of the rear panel. Others will use a perforated screen top-cover to allow heat to escape. The latter, incidentally, is not limited to cabinets with bottom mounted fans.

We often see equipment installed in standard 19-inch wide relay racks. Most of the principle manufacturers of relay racks also make dual squirrel cage blowers to cool the equipment. The blower is usually mounted at the bottom rear of the cabinet, and (sometimes) hoses, that resemble automobile radiator hoses, are used to distribute the forced air to the modules installed in the cabinet. A large exhaust port along the top edge (front or rear) will permit the heated air to escape. It is usually the practice to mount the worst heat generating modules closest to the top of the cabinet. Otherwise, rising heat from lower units tends to overheat the upper units.

Heat generated in any type of electronic equipment can be substantial. I can remember a 5000-watt (output) AM radio station that was wasting almost 6000 watts of anode heat dissipated by their water-cooled final amplifier and modulator tubes. They were able to reroute the heated water from the inside of the transmitter to external radiators, and it was sufficient to heat the transmitter building and the engineering department section of the main building. All winter long, even on the coldest days, it was usually too hot inside of the area by "waste" heat from the electronic equipment.

323

In another case, I can remember a medical computerized ICU/CCU patient monitoring system that averaged one expensive service call every 10 days or two weeks. The failure rate dropped to one every *three months* after blower fans were installed in the cabinet and ventilation holes were added.

POWER LINE PROBLEMS

Problems with AC power mains can affect almost any type of electronic equipment. My window air conditioner, for example, bears a label that in effect says the unit may burn out if operated during power company brownouts and line noise problems.

Brownouts occur when the AC power mains voltage drops from its specified range of 105 to 120 volts AC (rms) to 90 to 105 volts. Under these conditions, the DC voltage available from the rectifiers also drops. If it drops below the level required at the input of the DC voltage regulator circuits, then the voltage applied to the electronic circuits will be interrupted, even though there is DC applied to the regulator input.

It is never good to lose the DC supply voltage in any electronic equipment. In a digital computer, however, the problem is compounded by the fact that you may lose the data that is stored in volatile (semiconductor) memory. There are some circuits available that will guard against this problem, such as uninterruptible power supplies and special circuits that will dump memory contents to a disc when a power failure occurs. These usually operate from batteries on board the computer, or from the energy stored in the DC power-supply filter capacitors. Regardless, it is a sad state of affairs for the user when the machine goes down.

Another significant problem in AC power mains-operated equipment is that there is a considerable amount of noise present on the line. An oscilloscope analysis of many power lines will show frequent high-voltage transients and often a constant level of nonimpulse, high-frequency noise. Those nice AC sinewaves shown in our elementary electronics textbooks (including some by this author) actually belie the real situation. The truth is quite different. There are, according to one study, at least 10 transients on the average *residential* power line every day. Industrial and commercial lines can be expected to be even worse because of the types of equipment that may be operated on those lines.

The problems presented by having a trashy AC power mains source can be particularly upsetting to digital equipment. Noise spikes propagated through the power supply are seen a pulses by

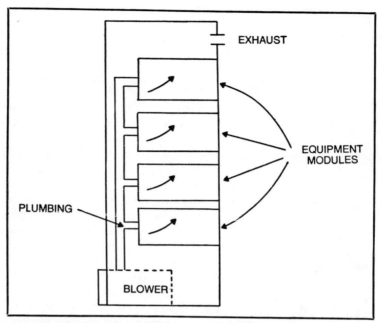

Fig. 21-3. Airflow distribution in rack mounted system.

the digital circuits. They are in effect indistinquishable from valid signals. They will set and reset flip-flops, increment or reset counter circuits, and either pass through or open or close logic gates at inopportune times. They are viewed as being just like regular, valid, pulses normally seen in the equipment. I once worked in a building that contained a large amount of electronic instrumentation that was used in medical and life sciences research. Much of this instrumentation was based on digital circuits, and there were more than a few computers, especially after microcomputers became easily available at low cost. The local power company gave a premium rate to those consumers who installed load-balancing equipment at the service entrance. Such equipment senses the current level drawn on each of the three phases. It then uses triac switches to disconnect some circuits from the more heavily loaded branches of the building and reconnects them to the more lightly loaded branches. It seems that power line efficiency is peak when all three available phases are loaded identically. That equipment represents a significant savings to the building management and the local power company. But it also throws high-voltage spikes down the lines inside the building, and those spikes

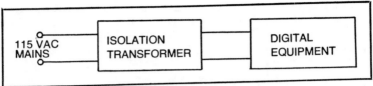

Fig. 21-4. Use of isolation transformer.

play heck with electronic equipment. One humorous aspect of the situation occurred early in the building's history, and amply illustrates the problems: All 125 freshman medical students flunked an examination that had been graded by computer. The exam was multiple choice, and the students marked IBM answer sheets. The answer sheets were then read by an optical scanner, which was connected by an interface cable to an IBM keypunch machine. These machines, working together, would read the answer sheets and then punch out one IBM Hollerith card for each student. These cards were then hand-fed into a card reader connected to the university's IBM 370 at the computer center across campus. The computer would grade the exams, rate the student's performance against others taking the examination, and do certain statistical chores for the professor. But in one case, the optical scanner was reading the answer sheets during a time when the power switches were sending transients down the line. The keypunch saw them as valid signals, and punched a lot of extra holes in the student's cards. You can imagine the uproar when 125 of the nation's future physicians saw the posted computer printout of their scores!

Power line conditioning, then, can become critical when digital equipment is used. Some cases of intermittent problems are so frequently due to power line problems, that the suppression practices that are next discussed are almost standard fare with digital equipment field service personnel.

In ideal situations, one would suppose that transients should be suppressed *at the source*. If the cause is known to be a certain motor, arcing contacts, or some similar source, then we can apply a capacitor, rf shielding, or an rf suppression filter, as needed.

But the situation facing most digital service personnel is a little different; the ideal is seldom realized. We usually find that the source is unknown, too multitudinous, or is impossible to suppress. We also find situations where it is merely uneconomical to suppress the noise at the source. It might be a lot cheaper to suppress the noise at the service entrance of the digital computer. In those cases, we will have to *condition* the AC line and the DC power supply.

One quick fix for many power source woes involves using a line voltage regulator/isolation transformer (Fig. 21-4) between the power mains and the equipment. In fact, some manufacturers either build such a transformer into their equipment, or include it among the site preparation specifications given to the customer. In the latter case, they may decline to honor their warranty if the transformer is not included. These transformers are models designed specifically to attenuate pulse transients on the line, and provide some regulaton of the AC voltage supplied to the equipment connected to the secondary. Both *Sola* and *Topaz* make various models for applications ranging from small desk top computers to large mainframe computer systems.

Another solution, one used by many manufacturers, is to provide rf filters at the point where the AC power enters the equipment. Such a filter circuit is shown in Fig. 21-5. The rf chokes are selected to have a negligible reactance at 60 Hz, but considerable reactance at radio frequencies. The capacitor values are selected to have a low rf reactance and a very high 60-Hz reactance—usualy 500 pF to 0.1 μF is specified. The entire filter assembly is mounted inside of a shielded enclosure, and will be located very near the service entrance to the equipment. Some manufacturers use chassis mounted AC power receptacles that have built-in, shielded rf filters. This means that no unfiltered AC power wiring will be inside the equipment cabinet.

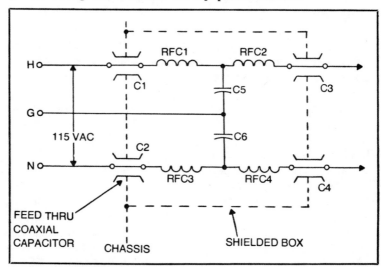

Fig. 21-5. Brute force rf filter.

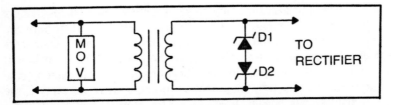

Fig. 21-6. Use of MOV and Zener diodes to suppress transients.

Two other approaches are shown in Fig. 21-6. The device shunting the primary winding of the transformer is a *metal oxide varistor* (MOV). General Electric is a prime manufacturer of these devices. The MOV is, in essence, a voltage sensitive resistor. If the voltage across the MOV remains at normal line values (170 volts peak), then nothing happens. The MOV has a high impedance under these conditions. But if a large transient comes along, the peak voltage will exceed the threshold, causing the MOV to exhibit a large drop in impedance. Because of the very short duration of the transient, the sharp drop in resistance across the AC line does not result in an appreciable increase in the total current drain. It will not blow the fuse.

The back-to-back Zener diodes across the secondry are used to clip transients. They will have a breakover voltage that is greater than normal peak AC values encountered in the circuit.

OTHER GLITCHES

A *glitch* can be defined as an unwanted pulse in a digital circuit. Glitches are caused by different things, but they all raise hob with circuits. They will do all of the nasty things attributed earlier to power line noise pulses, but are sometimes a lot harder to nail down.

Fig. 21-7 shows one cause of glitches. Pulses A and B are supposed to be coincident at the trailing edge. In situations where this coincidence is important, the actual time difference is not zero, but some small difference *t* (trace C). This can be propagated as a glitch, an unwanted and possibly harmful pulse.

The situation described above can be caused by any of several factors. One of these is the propagation delay of the digital IC devices being used. They normally require a few nanoseconds in order to pass an input change over to the output, and this causes a slight delay. If a particular IC or one particular manufacturer's IC is a little too slow or a little too fast, then problems occur. This is especially common when devices of two different manufacturers or two different logic families are mixed. But note that it can occur anytime. The solution sometimes is to change ICs until one is found that works. But sometimes the problem is due to a difference in a particular series IC, or those made by a specific vendor. In that case changing vendors, lot numbers, or series will help. The 74H, 74S and 74LS devices with similar suffix numbers are pin-for-pin compatible, yet have different operating spaces and propagation times.

In still other cases, where the problem seems to be related to timing, the equipment manufacturer will instruct the field service people to lengthen or shorten a pulse duration. On several occasions, for example, I have seen manufacturers change the RC time constants of one-shot multivibrators in order to eliminate a glitch problem.

Another type of glitch is caused by failure to properly bypass the DC power supply lines on the printed circuit boards. Analog electronic PC boards can often get away with very few bypass capacitors; i.e., some in the power supply and a couple on each PC card. But digital circuitry is a lot more critical in this respect. All of the fast, high-current drain, IC logic families are subject to this type of problem. TTL seems especially susceptible. The rapid pulse-like turn-on times seen in TTL circuits cause abrupt changes in the DC current drawn from the +5 VDC power supply. Fast rise time leading edges have a very high frequency component, so the distributed inductance and resistance of the +5VDC/ground traces on the P.C. board become important. When the TTL output goes LOW, an instantaneous increase in the current drain occurs. If the power supply were perfect, of course, then this would not have any affect and no problem would exist. But because the supply distribution wiring looks more like an RL network (Fig. 21-8), expect a sudden, local drop in supply voltage. The solution to the problem is to use wider (low-inductance) PC board lands for the +5V and

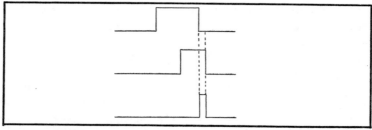

Fig. 21-7. Generation of a glitch.

ground paths and a large number of bypass capacitors distributed throughout the system.

Various protocols exist for how the bypass capacitors are distributed. Almost all of them want to see a 1 to 25 μF tantalum capacitor at the point where +5 VDC enters the PC board (usually right at the card-edge connector). Some also want to see a 25 to 250 μF/tantalum capacitor at the farthest extreme of the +5 VDC line, and a capacitor value of 100 μF/A at the output of the voltage regulator. The latter improves the transient response of the regulator, while the others improve the transient response of the PC board DC distribution system.

Almost all designers like to see an ample supply of small disc ceramic capacitors distributed all across the printed circuit board. These are of critical importance in many circuits. Some designers specify 0.001 μF/disc at the +5 VDC terminal of every TTL IC used in the circuit. Others use 0.001 to 0.1 μF every two or three TTL ICs. Regardless of which is preferred, some form of loyal bypassing must be supplied. Current stored in the capacitors can be dumped into the circuit when the protected IC suddenly increases its current demand (usually when the output goes LOW). This prevents the glitch and effectively cancels the effect of the distributed resistance and inductance of the line. Good bypassing will filter out most of the glitches that are created and propagated by the DC power system on any given PC board.

BUSES

Most digital equipment uses data or address buses. These are parallel signal lines used to form a digital word in parallel format. In a typical eight-bit microcomputer, for example, they will have an eight-bit *data bus* and usually a 16-bit address bus. For our example, let us consider the address bus. Most of these machines can support up to 65,536 (2^{16}) different eight-bit memory locations. This is due to the bit-length of the address bus; 16-bits can produce up to 65,536 different, unique, binary word combinations. In order to make a 16-bit bus, however, we need 16 parallel conductors!

The idea of a bus came into being to reduce the amount of wiring and circuitry needed in digital equipments. We can parallel a large amount of circuitry on a common bus. Decoders at each device, or circuit, are then used to allow response only to a unique code. Consider eight-bit memory locations normally used in digital computers. The eight-bit data bus is connected in parallel to all 65,536 different memory locations (or some submultiple, as is

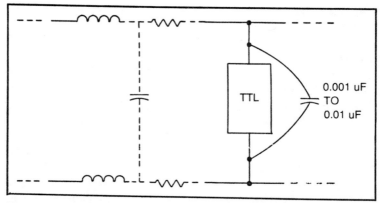

Fig. 21-8. Equivalent circuit for TTL power distribution.

common when 1K and up memory chips are used). Similarly, the 16-bit address bus is connected in parallel to all 65,535 sets of address lines. Any given memory location, however, will be effectively turned off unless it sees the unique code for its address on the address bus.

But buses can cause problems. No device has a perfect input, especially TTL devices. Every device that you connect to the line draws a small amount of DC current, even when it is officially turned off. When dozens (or dozens or dozens) of inputs are connected in parallel with a bus line, the voltage supposedly present in the HIGH condition is dragged down by ordinary voltage divider action. In some cases, the voltage reduction is so severe as to place the potential of the HIGH state in a gray region where TTL devices have difficulty determining HIGH from LOW.

One cure for this problem is to use *passive termination* of the bus lines (Fig. 21-9). We terminte each of the lines with a pull-up resistor to the +5 VDC source. When the line is LOW, then the bus end of the termination resistor is brought to ground and does not interefere. But when the line is HIGH, it sees a high impedance to ground and that makes the voltage on the bus wire closer to +5 VDC.

An *active termination* is essentially the same idea, but uses a small, independent, regulated power supply to accomplish the same mission. There is some indication that active termination is superior.

Ringing is another problem on buses, especially long buses. The bus signal lines are usually long (thereby increasing R and L), and quite close together (increasing C). Applying a pulse, then, is the same as ringing an equivalent RLC circuit. Following each pulse we see a damped sine wave oscillation. Termination, especially the active variety, decreases the duration of the oscillation to almost nothing.

RFI FROM DIGITAL EQUIPMENT

Digital circuits operate on pulses, so any digital equipment will usually have a considerable amount of pulse energy present all of the time. This is a situation that is ripe for the generation of radio frequency interference, or RFI. You can see an example of this phenomenon by holding a pocked calculator, especially a model using LED readouts, close to an AM radio receiver. You will hear a large amount of hash-like interference. My own home microcomputer has a "demonstration" program that uses RFI to play *The Star Spangled Banner* on a nearby AM radio receiver, while printing an asterisk and "x" pattern on the video terminal screen that resembles the US flag.

But putting games aside, RFI can be a serious problem. Digital equipment must often coexist with other equipment. Sensitive radio receivers, or RFI prone scientific instruments simply cannot operate properly with so much RFI trash floating around.

Part of the problem is radiation via the AC power line wiring. Fortunately, the same procedures that protected the digital equipment from trash already on the power line will also prevent transmission of the RFI from inside of the equipment. This is especially true of rf filters and isolation transformers.

But also be concerned with radiation from other than the power line. A prime source of radiation is interconnection cables, such as between the keyboard and the computer (although any interconnection cable can be the culprit). A 3 to 5 foot cable to a peripheral device is a very sufficient transmitting antenna.

Radiation also comes from the circuitry inside of the cabinet. This is a phenomenon long familiar to communications people. Shielding of the offending circuitry is often the best solution for both types of radiation. Any book on RFI problems should be more than sufficient to provide troubleshooting ideas. Some of the best, incidentally, are found in the amateur radio sections of book catalogs.

UNINTERRUPTIBLE POWER SYSTEMS

An uninterruptible power system, or UPS, is often specified for computers that *must* operate all of the time. Any computer that

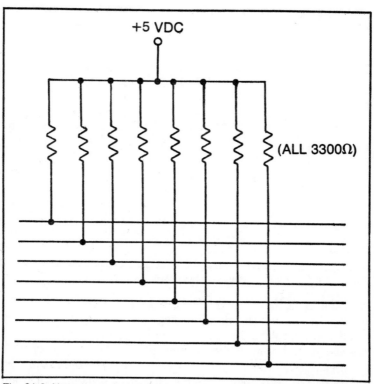

Fig. 21-9. Use of pull-up resistors in a passive termination of the data bus.

contains critical, real-time data is a candidate for such a system. As the name implies, a UPS cannot allow the power system to be interrupted. There are several ways to accomplish this feat. One is to use batteries to power the computer. The AC power system is used to charge the batteries. If loss of the AC power occurs, a control circuit turns on an emergency generator system to continue charging the batteries. The batteries will contain sufficient energy to operate the computer for the short period of time when the gasoline, or diesel, engine driving the generator is starting up.

Another UPS system is shown in Fig. 21-10. This technique also uses an engine-driven generator, but the generator is always used to operate the computer. Normally, the engine is off, and an AC motor turns the AC generator shaft. The output of the AC generator is then used to power the computer. A line voltage sense circuit determines when the AC line power is interrupted, and

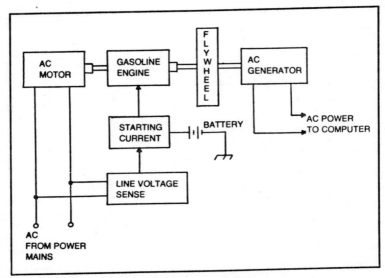

Fig. 21-10. One type of uninterruptible power system.

turns on an engine-starting circuit. When the engine comes to full speeed, it will be capable of turning the AC generator shaft by itself. But what about the time between loss of the mains power and full-speed operation of the engine? This is taken care of by a *flywheel*! The flywheel is normally kept turning by the AC motor. By its nature, a flywheel does not come to a grinding halt when drive power is lost. There is enough energy stored in the flywheel to keep it running for a long time. This gives the engine sufficient time to start.

Chapter 22
Other Equipment

Many years ago, when I was still fresh from a high school electronics shop course, I fancied myself as a hot-shot radio/TV repairer. One of the first jobs given to me when I reported to my first *real* job was to—ichhh—clean and lube a record player. The point is that servicers of exotic electronics equipment often must wear multiple hats. They often service a range of products that include electrical, mechanical, and electromechanical devices. Digital servicers have the same problem, but some of their electromechanical junk seems a little more exotic than an old RCA 45RPM phonograph; in fact, the equipment can be more complicated than the computer or device it serves.

PRINTERS

Various types of printers are used to provide hard copy output for computers and other digital instruments. Some printers are little more than a drop-in OEM module that has been placed in a side corner of the front panel of the main instrument. Other printers are teletypewriters (Teletype, IBM Selectric, etc.), while still others are mechanical wonders that print whole lines at a time. The actual printer used is determined by the application and selections of the designer. There are, however, only a few technologies used. These can be broken into two basic categories: *impact* and *nonimpact*.

Impact printers, much like an ordinary typewriter, use a hammer or something similar to strike a printhead of some sort. Nonimpact printers use thermal or electrical techniques to deposit the letters or numbers on the paper.

One of the simplest printers is the numerical printer of Fig. 22-1. This type of printer is frequently seen in digital instruments

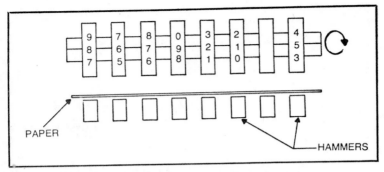

Fig. 22-1. Numerical printer using a mechanical counter system.

that output a number or value. It is based on the simple mechanical counter (Veedor-Root mechanism). Such a counter is advanced one place for each pulse applied to its electromagnet. Note that these are not electronic counters, but electromechanical. They are like the little hand-operated counter you may have seen. Each time a button is pressed, the counter increments by one. In this case, however, the counter "button" is the armature of a relay or solenoid. Each time the relay is energized, the counter advances by one. The counter operates by having the digital instrument output the number of pulses that corresponds to the desired count. The counter will advance to that point, at which time a set of solenoid operated strike hammers impact on the counter wheels. A piece of specially treated paper is placed between the hammers and the print head, so the letters are left on the paper. This technique is used only in the simplest, lowest-cost printers. They are quite popular, however, because many instruments only need to output a numerical value. In many cases, the printer will print many lines because it will advance the paper after each print and make a fresh impression using new data.

One of the earliest types of printers used with computers was the old-fashioned Teletype® machine. Although there have been several models over the years since World War II, the one that has proven most popular with computer manufacturers is probably the Model 33. This machine is able to respond to ASCII code, instead of Baudot code, and is therefore more compatible with computer standards. The Model 33 uses a print mechanism such as shown in Fig. 22-2. The characters are embossed on a metal cylinder that is mounted upright with respect to the paper. The mechanism senses the ASCII code and mechanically decodes it. This operation causes the cylinder to rotate to the position containing that letter, and then

336

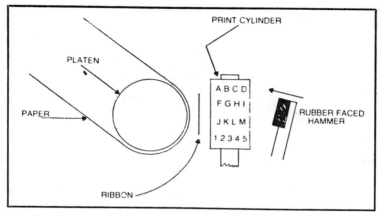

Fig. 22-2. Print cylinder such as used in teletypewriters.

the hammer strikes the cylinder. A ribbon is placed between the cylinder and paper, so an impression is left on the paper. The teletypewriter operation is more like that of an ordinary typewriter in that respect.

Very popular for many years has been the IBM *Selectric*® typewriter mechanism. This mechanism is the so-called "golf ball," or "flying ball" system (Fig. 22-3). The balls are interchangeable for different type fonts. The Selectric is an office typewriter, but for years IBM offered a Selectric I/O model for computers. Even today, while IBM does not make an I/O model anymore, a number of vendors buy Selectric mechanisms from IBM and build them into their own cases. The secondary manufacturer will perform the modifications to make the Selectric computer-compatible and then sell it to the user. An example of these are the Andersen-Jacobsen terminals.

Fig. 22-3. IBM Selectric "flying ball" printer.

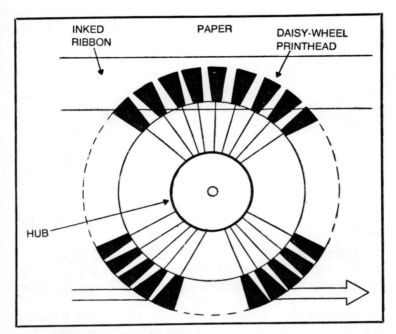

Fig. 22-4. Daisy wheel printhead.

Another printer mechanism is the daisy wheel of Fig. 22-4. In this type of machine, the characters are embossed on pads at the ends of the spokes of a wheel. The wheel rotates constantly, and when the desired character is opposite the print space, the hammer strikes the pad, making the impression on the paper. Neither the cylinder (Teletype), flying ball, nor the daisy wheel need special paper, as they use inked ribbon to make the impression.

One other type of impact printer is the dot-matrix impact print head of Fig. 22-5. In this system, the characters are formed in a 5×7 dot matrix by little pins emanating from the print head. The decoder circuitry will cause the correct pins needed for the character to stick out of the head. These pins impact the paper and leave the impression. Either ribbon or specially treated paper can be used.

Some high-speed printers use a belt system, such as shown in Fig. 22-6. The characters are embossed on a belt that is constantly rotated. When the desired character is opposite of the print space, a hammer strikes and leaves the impression.

Most thermal and electrical printers use the 5×7 dot matrix system, but the dots are formed in a manner different from the

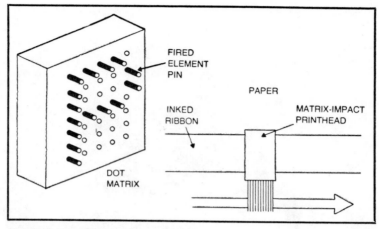

Fig. 22-5. Dot matrix impact printhead.

impact version. In the thermal type, each dot is a hairpin loop of wire that is heated to incandescence when the dot is needed to form a character. In the electrical type, the dot forms one terminal of an electrical circuit. The platen of the printer forms the other contact. When a current passes through the wire, the connection between print and platen is made through electrically conductive special paper. A black dot is left at the point of contact. Both thermal and electrical printers are very popular; we will, however, see far more of the thermal types.

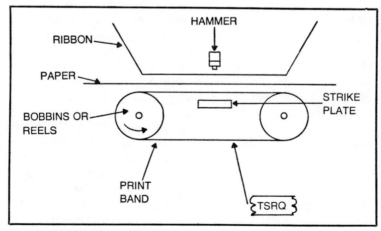

Fig. 22-6. Print band system.

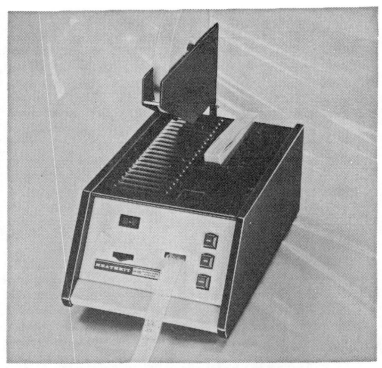

Fig. 22-7. Paper tape reader/puncher (courtesy of Heath Co.).

PAPER TAPE READERS

Figure 22-7 and 22-8 show a paper tape reader; this model is by Heath. Paper tape was once used extensively in teletypewriter work to transmit repeated messages or to transmit high-speed messages. In the latter case, several operators would cut the messge onto tape. Then a high-speed teletypewriter would transmit it at a rate faster than could reasonably be followed by a human operator. Paper tape was also used extensively as a means for program storage, especially in the earlydays of computer technology. It is still used widely, however, even though it has been largely eclipsed by magnetic tape and magnetic discs. Most tape machines will both read and punch tapes.

Figures 22-9 and 22-10 show two methods by which paper tape can be read. The older method used sense wires or pins to make electrical contact with a grounded plate underneath the tape. If the bit at that location was a mark (a1), then there would be a hole in the tape at that location and the sensor would make contact. But

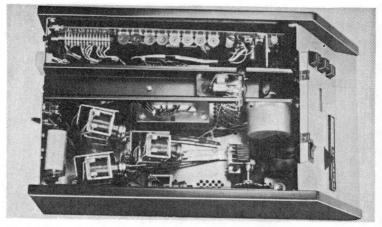

Fig. 22-8. Internal view of the Heath paper tape reader/puncher.

if the bit was a space, then there would be no hole so the sensor would not make contact with the plate because paper is an insulator.

In more modern equipment, however, an optical method does the same job. An array of photocells or phototransistors takes the place of the grounded plate, while light-emitting diodes (LEDs) take the place of the sense wires. If there is a hole in the paper (MARK), then light from the LED shines on the photocell. But if there is no hole in the paper, light cannot reach the paper. The optical method is also used in nonpunched paper tape in which inked dots are used to represent the bits. Card readers use essentially the same technology to read the data punched into the card.

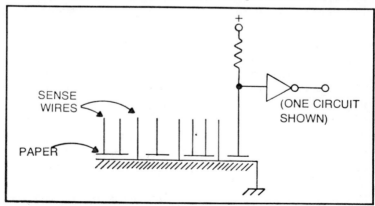

Fig. 22-9. Electrical sense wires for reading paper tape and punch cards.

341

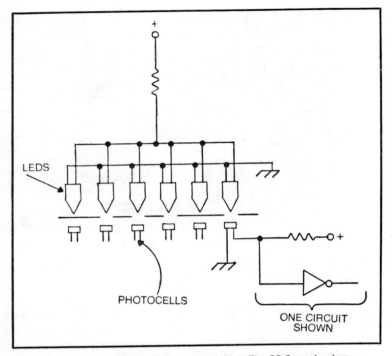

Fig. 22-10. Optical reader does the same job as Fig. 22-9 mechanism.

MAGNETIC STORAGE DEVICES

All computers have a limited on-board or internal memory. In older computers, this memory was magnetic cores, so the on-board memory was called *core*. Today, solid-state memory is used. The total internal memory capability of a computer is determined not by the *physical* size of the memory (solid-state is generally smaller than core), but by the length of the address bus. In a typical microcomputer, the address bus is 16 bits in length, so the computer can support up to 2^{16}, or 65,536, different memory locations. Many applications require far more memory, however, instead of building monster computers with megabit address buses, though other forms of temporary or semipermanent storage are used. The on-board memory can be loaded with blocks of data from the external memory as needed. Alternatively, the computer can write data from the internal memory to the external memory device. There are several different types of external memory: magnetic tape, magnetic drums, and magnetic discs. The last type is further subdivided into hard discs and floppy discs.

Two types of magnetic tape system are in use: audio and digital. The audio tape system is slow (less than 1100 baud) and popular only among the low-cost end of the home computer market. Few, if any, professional or business systems will use audio tape. In these systems, the bits are represented by two tones, one for HIGH and another for LOW. In my own Digital Group system, for example, the HIGH condition is represented by 2125 Hz and LOW is represented by 2975 Hz. The decoding method is similar to those discussed in Chapter 14.

There are only a few advantages to audio tape systems. One is low cost. The tape system used is ordinary audio cassettes. In actual practice, any good quality audio cassette machine is usable, and the tapes are the so-called middle-quality. As a result, program entry and storage are made available to low-end machines. But audio cassettes are also full of problems. Even at slow data rates there are frequent misreads. At the 1100 baud rate of the Digital Group TVC interface, read errors are far more common unless very good quality machines and tapes are used. Tape speed is also very critical in audio cassette systems. Recall from Chapter 14 that tone decoders tend to use filters to separate the HIGH and LOW tones. If the speed is in error, then the tones received from the recorder will be outside of the passband of the filter (or alternatively, the phase-locked loop capture range). But you will see audio cassette interfaces in most home computers, where cost is a determining factor in design choices.

Digital tape uses special magnetic oxides that can be magnetized by passing a pulse current through the record head. The bit

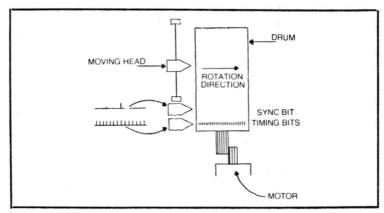

Fig. 22-11. Magnetic drum memory.

Fig. 22-12. IMSAI dual disc drive.

is represented by the magnetic state at the correct point on the tape. An example of digital magnetic tape machines is the Phi-Deck machine. Many users of digital cassette machines use ordinary audio-grade cassettes, but unlike audio cassette systems, must use only the highest grade of cassette. Some manufacturers offer certified computer-grade cassettes for these machines.

Large systems use reel-to-reel tape drives that are very high quality machines. In many of these machines, the tape is placed against the head more firmly by vacuum pressure. These machines are large drives and very costly, so they are found only in larger commercial computers.

An example of a magnetic drum memory is shown in Fig. 22-11. The drum is a cylinder whose surface is coated with a magnetic material similar to that used in magnetic tape (a ferric oxide). A read head is positioned so that it could move up and down in the vertical plane; then any point along that line can be read. The drum rotates, so by knowing where on the vertical plane and where on the rotation cycle a bit is located, any given data can be found. In order to know where the drum is in the rotation cycle, some timing bits must be produced. These bits are usually along one edge of the drum and are read by a fixed head. Many timing bits are received each cycle, and they can be electronically counted. One additional bit, a synchronization bit, is needed. There will be only one sync bit per rotation. If this bit is used to reset the timing counter to zero once each revolution, then the state of the counter, as determined by the timing bits, will locate the position of the drum for the computer.

Fig. 22-13. Digital Group dual disc drive.

Discs are also magnetic devices. The hard disc is coated with ferric oxide and resembles an oversized phonograph record. Hard discs are used in minicomputers and larger computers. In smaller systems based on microcomputers, however, floppy discs are used. These are similar in size to the old 45 rpm phonograph records. Both 5-inch and 8-inch diameter models are available. They will not operate at the same speed as the hard discs and will not store nearly as much data, but they are a low-cost alternative to mass storage. An example of a disc drive is shown in Figs. 22-12 and 22-13.

PLOTTERS/RECORDERS

One last class of device which we will consider is chart recorders and plotters. Examples of plotters are shown in Figs. 22-14 through 22-16. Most of these are analog instruments, but some are digitally operated. Analog readers are oscillographic, and will reproduce the waveform of the applied signal onto graph paper. An analog type recorder must be driven by an analog signal, usually a voltage. If we are to drive such an instrument with a computer, then we must first convert the binary output of the computer to a

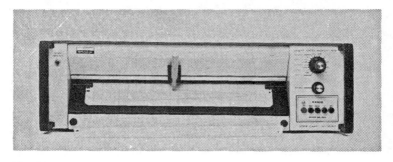

Fig. 22-14. A chart recorder.

voltage. This is done by a digital-to-analog converter (DAC), as described in Chapter 18.

Several types of recorders are available. Fig 22-14 shows a Y-time recorder. The paper is in strip form, and is sprocket driven—a sprocket fits into holes along the edge of the paper. A pen makes a mark on the paper proportional to the amplitude of the applied signal. If this amplitude is sketched while the paper is moving, the waveform is traced onto the paper.

Another type of recorder is the X-Y recorder in Fig. 22-15. This type of recorder uses ordinary graph paper, often held in place by a small vacuum applied through the paper plate. The pen is moved vertically and horizontally by a pair of servomechanisms. The pen position will be proportional to the applied potential. X-Y servo-recorders are more fully described in TAB Book No. 930, *Servicing Medical & Bioelectronic Equipment.*

A more sophisticated form of X-Y recorder is the graphics plotter of Fig. 22-16. This model is capable of making complex patterns on the paper by using a pen lift system. The pen is lifted off of the paper under command from the computer and then set back down where a plot is needed.

Analog recorders use either permanent magnet moving coil (PMMC) galvanometers that are much like D'Arsonval meter movements, or the X-Y servomechanism. But digital recorders use

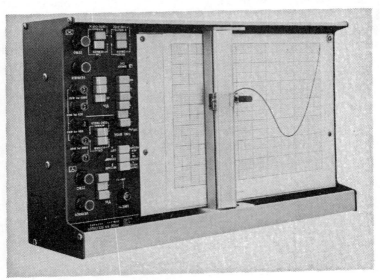

Fig. 22-15. Another chart recorder.

Fig. 22-16. Hewlett-Packard graphic plotter model 7203A.

stepper motors to drive the pen (one each for vertical and horizontal). A stepper motor will advance only when a pulse is received. A constant level (either HIGH or LOW) will not cause the motor to advance; only a *pulse* will do the trick. When these devices are used, the stepper motor for each axis is connected to one bit of an output port or to a serial output port. These devices are analogous to *mechanical DACs*, with the analog output being pen position.

Appendix A

Logic Level Detector Circuits

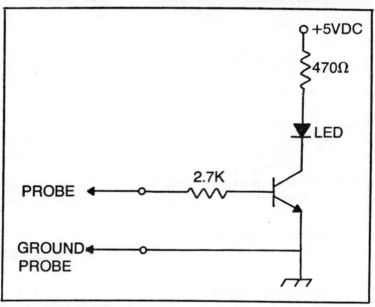

Fig. A-1. Logic level detector circuit.

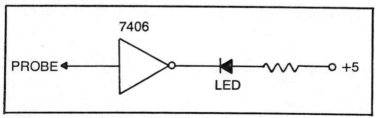

Fig. A-2. Logic level detector circuit.

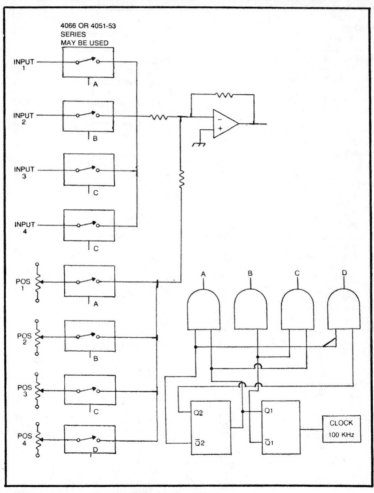

Fig. B-1. A four-channel oscilloscope switch.

Index

A

Active termination 331
ADC 259
Adder circuits 89
    full-adder 89
    half-adder 89
Alphanumeric codes 38
ALU 223
American Standard Code
    for Information Interchangerd 40
Amplitude 12
Analyzers 316
AND gates 73
Analog signal 12
    continuous 12
Analog-to-digital conversion 269
    converters 259
Arithmetic logic unit 223
ASCII 38,40
Astable multivibrator 152
Asynchronous peripheral 189

B

Base-16 counter 122
Baud 191
Baudot code 38
BCD 36
     124
Binary addition 25
    arithmetic 24
    coded decimal 36
    coded decimal 124
    multiplication 29
    numbers 18
    number system 11, 15
    readout, N-bit 135
    readout, simple 134
    ripple-carry counter 120
    subtraction 25
    weighted resistor ladder 263
    word decoders 235
Bistable multivibrator 152
Bit 18
Buses 330
Byte 18

C

Carry-one output 90
Cathode ray tube video terminals 10
Central processor unit 215
Circuits, adder 89
    DAC 263
    I/O select 240
    power-supply 308
    serial data 190
    subtractor 92
    teletypewriter 195
Clocks 153
Clock generator 152
CMOS 56-64

    handling rules 64
CPU 215
CRT video terminals 10
Commercial electronics servicing 117
Communications channels 173
Comparator 154
Complementary metal oxide
    semiconductor logic 56-64
Complementors 68
Computers 217
Control logic section 223
Counters 119-133
Current sink 47
    source 47

D

DAC 269, 259
    circuits 263
Data converter resolution 261
DC power supplies,
    basic principles 287
Decade counters 123
    displays, simple 135
Decimal counters 122
    numbers 16
    number system 16
Decoders 235, 238, 141-144
Demultiplexer 180
Device selector 234
Digital counter circuits,
    types 119
    electronics 11
    signal 12
    to-analog converters 259
    unmasked 14
    selector 177
Display multiplexing 144
Diode-transistor logic 49
Domain 12
Down counters 126
DTL 49
Duty cycle 160
    factor 160
Dual-slope integration 271

E

EBCDIC code 40
ECL 53
Eight-bit word decoders 238

Emitter-coupled logic 53
End-of-conversion 274
EOC 274
Excess-3 code 37
Exclusive-OR gates 79
Extended Binary Coded
    Decimal Interchange
    Code 40